Acceptance-Rejection Sampling a[ ] dimensional Monte Carlo Integrations Utilizing Mathematica®

Sujaul Chowdhury

AMERICAN ACADEMIC PRESS

AMERICAN ACADEMIC PRESS

By AMERICAN ACADEMIC PRESS

201 Main Street

Salt Lake City

UT 84111 USA

Email manu@AcademicPress.us

Visit us at http://www.AcademicPress.us

ISBN: 979-8-3370-8929-4

Distributed to the trade by National Book Network Suite 200, 4501 Forbes Boulevard, Lanham, MD 20706

10 9 8 7 6 5 4 3 2 1

Practical demonstrations of evaluation of multi-dimensional or high-dimensional definite integrals are *not* available. This book provides hands-on demonstrations of evaluation of multi-dimensional or high-dimensional definite integrals using acceptance-rejection sampling and user friendly program Mathematica®.

We have chosen suitable *multi-variable* probability density functions and obtained corresponding random variates using the random sampling method called *acceptance-rejection* method. We have performed symbolic computations using programs written in Mathematica® by which we performed the variate sampling and evaluated the integrals. Uses of different parts of the programs written in Mathematica have been narrated.

We have evaluated a number of multi-dimensional definite integrals using Monte Carlo method. These are 2, 3, 5, 7 and 10 dimensional definite integrals. The integrals are:

$$I_1 = \int_a^b \int_a^b (x_1 + x_2)^y \, dx_1 \, dx_2$$

$$I_2 = \int_a^b \int_a^b (x_1 + x_2)^{-y} \, dx_1 \, dx_2$$

$$I_3 = \int_{x_1=a}^b \int_{x_2=g}^h (x_1 + x_2)^{-y} \, dx_1 \, dx_2$$

$$I_4 = \int_a^b \int_a^b \int_a^b (x_1 + x_2 + x_3)^y \, dx_1 \, dx_2 \, dx_3$$

$$I_5 = \int_a^b \int_a^b \int_a^b (x_1 + x_2 + x_3)^{-y} \, dx_1 \, dx_2 \, dx_3$$

$$I_6 = \int_a^b \int_a^b \int_a^b \left(\frac{x_1}{x_2 + x_3}\right)^y \, dx_1 \, dx_2 \, dx_3$$

$$I_7 = \int_a^b \int_a^b \int_a^b \left(\frac{x_1}{x_2 + x_3}\right)^{-y} \, dx_1 \, dx_2 \, dx_3$$

$$I_8 = \int_a^b \int_a^b \int_a^b \int_a^b \int_a^b (x_1 + x_2 + x_3 + x_4 + x_5)^y \, dx_1 \, dx_2 \, dx_3 \, dx_4 \, dx_5$$

$$I_9 = \int_a^b \int_a^b \int_a^b \int_a^b \int_a^b (x_1 + x_2 + x_3 + x_4 + x_5)^{-y} \, dx_1 \, dx_2 \, dx_3 \, dx_4 \, dx_5$$

$$I_{10} = \int_a^b \int_a^b \int_a^b \int_a^b \int_a^b \left(\frac{x_1 + x_2}{x_3 + x_4 + x_5}\right)^y \, dx_1 \, dx_2 \, dx_3 \, dx_4 \, dx_5$$

$$I_{11} = \int_a^b \int_a^b \int_g^h \int_g^h \int_g^h (x_1 + x_2 + x_3 + x_4 + x_5)^y \, dx_1 \, dx_2 \, dx_3 \, dx_4 \, dx_5$$

$$I_{12} = \int_a^b \int_a^b \int_p^q \int_p^q \int_p^q \left(\frac{x_1 + x_2}{x_3 + x_4 + x_5}\right)^{-y} \, dx_1 \, dx_2 \, dx_3 \, dx_4 \, dx_5$$

$$I_{13} = \int_a^b \int_a^b \int_a^b \int_a^b \int_a^b \int_a^b \int_a^b (x_1 + x_2 + x_3 + x_4 + x_5 + x_6 + x_7)^y \, dx_1 \, dx_2 \, dx_3 \, dx_4 \, dx_5 \, dx_6 \, dx_7$$

$$I_{14} = \int_a^b \int_a^b \int_a^b \int_a^b \int_a^b \int_a^b \int_a^b \left(\frac{x_1 + x_2 + x_3}{x_4 + x_5 + x_6 + x_7}\right)^y \, dx_1 \, dx_2 \, dx_3 \, dx_4 \, dx_5 \, dx_6 \, dx_7$$

$$I_{15} = \int_a^b \int_a^b \int_a^b \int_a^b \int_a^b \int_a^b \int_a^b \int_a^b \int_a^b \int_a^b (x_1 + x_2 + x_3 + x_4 + x_5 + x_6 + x_7 + x_8 + x_9 + x_{10})^y$$

$$dx_1\ dx_2\ dx_3\ dx_4\ dx_5\ dx_6\ dx_7\ dx_8\ dx_9\ dx_{10}$$

$$I_{16} = \int_a^b \int_a^b \int_a^b \int_a^b \int_a^b \int_g^h \int_g^h \int_g^h \int_g^h \int_g^h (x_1 + x_2 + x_3 + x_4 + x_5 + x_6 + x_7 + x_8 + x_9 + x_{10})^y$$

$$dx_1\ dx_2\ dx_3\ dx_4\ dx_5\ dx_6\ dx_7\ dx_8\ dx_9\ dx_{10}$$

Here y is positive constant and a, b, g, h, p, q are also constants.

We have also evaluated the same definite integrals using *NIntegrate* command of Mathematica. % error between results obtained using Monte Carlo method and those obtained using *NIntegrate* command of Mathematica are obtained as check of Monte Carlo results.

Key words for the book are: numerical, multi-dimensional, high-dimensional, definite integrals, Monte Carlo Integration, Mathematica®, *NIntegrate*, inverse transform sampling, acceptance-rejection sampling, multi-variable probability density function.

This book will prove useful to graduate students for the course and Lab titled *Computational Physics* or *Computational Mathematics*. There are 6 chapters besides concluding remarks, reference and list of books for further use. The 1st chapter contains introduction to topics that are necessary to enable readers to assimilate remaining chapters of this book. Both inverse transform and acceptance-rejection sampling methods are introduced. Each of the remaining chapters deals with Monte Carlo evaluation of multi-dimensional definite integrals using random variates sampled by acceptance-rejection method.

This book is self-contained. Necessary introduction is provided in chapter 1. The book will contribute its mite in further evaluation of multi-dimensional Monte Carlo integrations. This book is 12th attempt of the author to build course books for the course and Lab titled *Computational Physics* or *Computational Mathematics*. Other books by the author in the series are listed below reference.

Sujaul Chowdhury
Sylhet, Bangladesh, 2025

content

Acceptance-Rejection Sampling and Multi-dimensional Monte Carlo Integrations Utilizing Mathematica®

Sujaul Chowdhury   www.sust.edu

Chapter I

Introduction

This chapter introduces readers to topics that are necessary to understand remaining chapters of this book. Both inverse transform and acceptance-rejection methods for sampling random variates are introduced. Use of acceptance-rejection sampling in case of multi-variable probability density function is also included. Variance reduction and importance sampling are also introduced. A technique of evaluating multi-dimensional Monte Carlo integrals to be used in subsequent chapters is described.

1.1 Random variable, continuous and uniform

In the context of Monte Carlo methods, the term *random variable* does not correspond to colloquial use of the term. We must indicate the values that the variable can assume and probabilities of occurrence of these values. We do not know value of the variable in any given case; but we know the values that the variable can take on and the probabilities of occurrence of these values.

We call a random variable x *continuous* if it can assume any fractional value in a certain interval, say a to b. If $p(x)$ is the associated probability density function, we have $p(x) \geq 0$ for $a \leq x \leq b$, the product $p(x)dx$ is probability that value of x lies in the interval x to $x + dx$, $\int_{a'}^{b'} p(x)dx$ is probability that $a' \leq x \leq b'$ where $a < a'$ and $b' < b$, and

$\int_{a}^{b} p(x)dx = 1$ if $p(x)$ is normalized.

A random variable is called *uniform* if $p(x)$ is a constant say C. If the interval for uniform random variable is 0 to 1, normalization requires $\int_{0}^{1} C \, dx = 1$ which gives $C = 1$. Thus normalized probability density function of uniform random variable in the interval 0 to 1 is $p(x) = 1$. We then denote the random variable by u. If the interval for uniform random variable is a to b, normalization requires $\int_{a}^{b} C \, dx = 1$ which gives $C = 1 / (b - a)$. Thus normalized probability density function of uniform random variable in the interval a to b is

$$p(x) = 1 / (b - a) \qquad\qquad \text{-------(1.1)}$$

We then denote the random variable by U.

Program number 1.1 provides us with 100 values of uniform random number u in the interval 0 to 1. See Table 1.1. The SeedRandom[n] command with a chosen value of n provides us same set of random numbers reproducibly. The value of n has been chosen such that quality of the random numbers is good. Here good means the random numbers are almost uniformly spread in the interval 0 to 1 as Figure 1.1 reveals. If we randomly pick up one of these 100 values of u, probability that it will lie in the interval 0.3 to 0.7 is $(0.7-0.3)/1 = 0.4$ which is equal to the interval itself.

Program number 1.1

```
N1=100

n=654321;

SeedRandom[n];

Table[{i=i+1,u[i]=RandomReal[]},{i,0,N1-1,1}];

TableForm[%,TableSpacing->{2,2},

TableHeadings->{None,{"i","u[i]"}}]

ListPlot[Table[{u[i],1},{i,0,N1-1,1}],Frame->True,

FrameLabel->{"u [i]","1"}]
```

Table 1.1 Showing 100 uniform random numbers u_i's (in the interval 0 to 1) obtained using program number 1.1 in Mathematica.

i	u_i	i	u_i	i	u_i	i	u_i
1	0.8618	26	0.9276	51	0.7833	76	0.8258
2	0.4287	27	0.0504	52	0.7877	77	0.4101
3	0.8596	28	0.3510	53	0.7012	78	0.6742
4	0.0342	29	0.1510	54	0.5937	79	0.2088
5	0.7597	30	0.1761	55	0.6616	80	0.8631
6	0.4609	31	0.3796	56	0.5202	81	0.2042
7	0.4189	32	0.5239	57	0.9024	82	0.3027
8	0.4746	33	0.3586	58	0.0050	83	0.0504
9	0.3799	34	0.4202	59	0.2183	84	0.8463
10	0.1687	35	0.5641	60	0.2361	85	0.4134
11	0.1899	36	0.1372	61	0.4477	86	0.2451
12	0.0364	37	0.0774	62	0.7308	87	0.5565
13	0.3335	38	0.8924	63	0.3429	88	0.0084
14	0.4257	39	0.9848	64	0.2260	89	0.3532
15	0.3860	40	0.4854	65	0.1079	90	0.7547
16	0.2445	41	0.3080	66	0.4879	91	0.1188
17	0.5881	42	0.4945	67	0.1655	92	0.5403
18	0.2322	43	0.9390	68	0.1218	93	0.5235
19	0.6743	44	0.5923	69	0.3544	94	0.8228
20	0.9089	45	0.1921	70	0.3523	95	0.5526
21	0.8759	46	0.3830	71	0.3895	96	0.8279
22	0.3658	47	0.6754	72	0.0911	97	0.6770
23	0.2945	48	0.2944	73	0.9631	98	0.7266
24	0.6719	49	0.9737	74	0.5184	99	0.4398
25	0.9104	50	0.4501	75	0.6328	100	0.2243

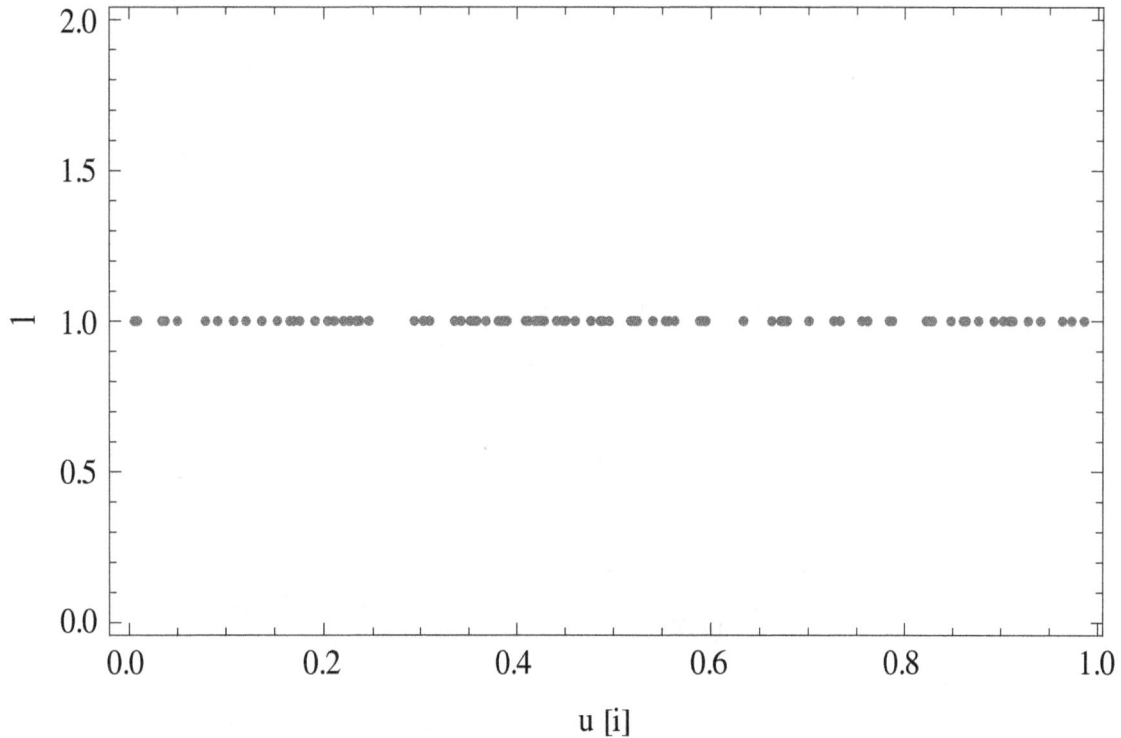

Figure 1.1 Uniform random numbers u_i's of Table 1.1 plotted against a stagnant integer 1 using program number 1.1. The 100 values of u are indeed almost uniformly distributed or spread in the interval 0 to 1.

1.2 Generating random variates by inverse transform method

Values of random variable η corresponding to a given probability density function $p(x)$ defined for the interval $a \leq x \leq b$ can be obtained by solving equation (1.2):

$$\int_a^\eta p(x)dx = u \qquad \qquad \text{------(1.2)}$$

where u is uniform random variable distributed in the interval $0 \leq x \leq 1$; see Table 1.1 and Figure 1.1. We now verify this.

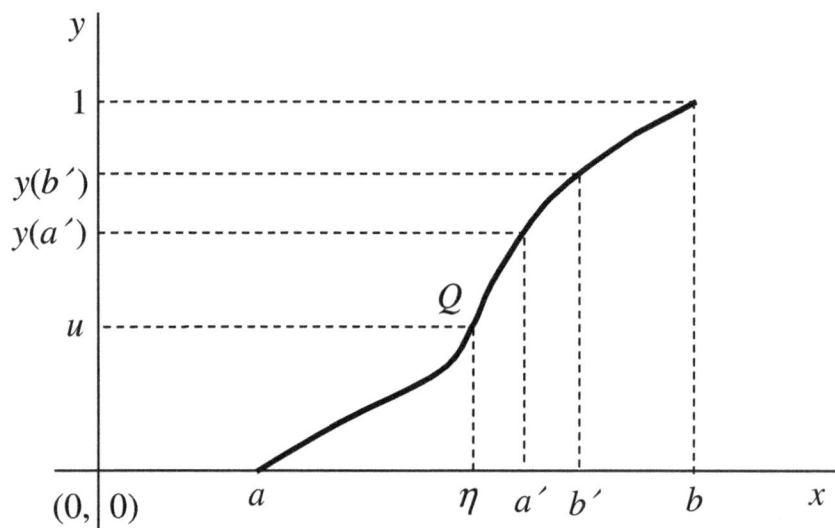

Figure 1.2 Figure to help verify inverse transform method of obtaining random variates.

Let $$y(x) = \int_a^x p(x)dx \qquad \text{------(1.3)}$$

Since $p(x) > 0$, and $\int_a^b p(x)dx = 1$, we have

$$y(a) = \int_a^a p(x)dx = 0$$

$$y(b) = \int_a^b p(x)dx = 1 \text{ emphasizing that } p(x) \text{ is normalized,}$$

and $$y'(x) = p(x) > 0.$$

This means that y monotonically increases from 0 to 1 for the interval $a < x < b$. See Figure 1.2. Straight line $y = u$ where $0 < u < 1$ intersects the curve $y(x)$ at Q giving us the value of η which satisfies equation (1.2): $\int_a^\eta p(x)dx = u$.

$$p\,(a' < \eta < b') = p\,(\,y(a') < u < y(b')\,) = y(b') - y(a') = \int_a^{b'} p(x)dx - \int_a^{a'} p(x)dx = \int_{a'}^{b'} p(x)dx$$

Thus random variable η which obeys equation (1.2): $\int_a^\eta p(x)dx = u$ has the probability density function $p(x)$. We need to evaluate this integral and obtain values of η for different values of uniform random variable u.

We now turn to a good example of obtaining random variates using inverse transform method. Values of random variable η *uniformly* distributed in the interval a to b are obtainable from equation (1.2): $\int_a^\eta p(x)dx = u$ with probability density function $p(x)$ given by equation (1.1): $p(x) = 1 / (b - a)$. As such we get

$$\eta = a + (b - a)\,u \qquad \text{-------(1.4)}$$

This is a very useful formula. If the probability density function $p(x)$ is such that the integral $\int_a^\eta p(x)dx$ cannot be analytically evaluated, we cannot or do not use inverse transform method.

If we wish to obtain random variates U uniformly distributed in the interval 3 to 5, for example, we have $a = 3$, $b = 5$ in equation (1.4): $U = a + (b - a)\,u$. Here u is uniform random number in the interval 0 to 1. As such $U = 3 + (5-3)\,u$. Using program number 1.2, we can obtain the 100 random variates U shown in Table 1.2. That U's are uniformly distributed in the interval 3 to 5 is evident from Figure 1.3.

Program number 1.2

```
N1=100;

n=654321;

SeedRandom[n];

Table[{i=i+1,u[i]=RandomReal[],U[i]=3+(5-3)u[i]},{i,0,N1-1,1}];

TableForm[%,TableSpacing->{2,2},

TableHeadings->{None,{"i","u[i]","U[i]"}}]

ListPlot[Table[{U[i],1},{i,0,N1-1,1}],Frame->True,

FrameLabel->{"U[i]","1"}]
```

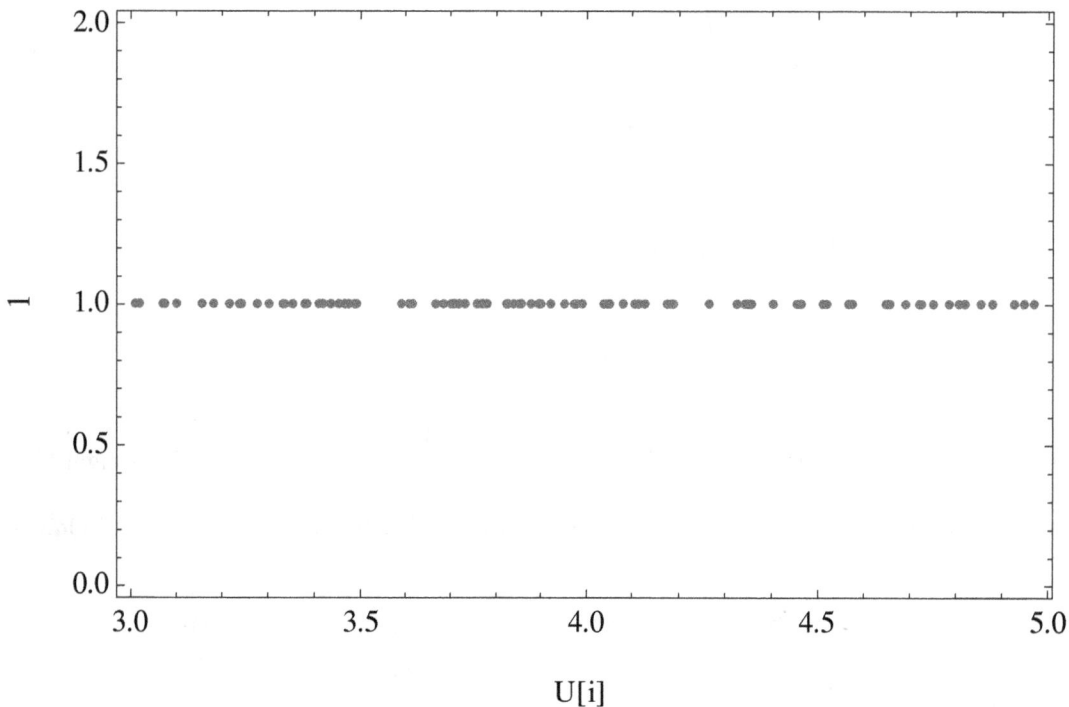

Figure 1.3 Showing that the 100 values of the random variate U are nearly uniformly distributed in the interval 3 to 5. Obtained using program number 1.2. Compare with Figure 1.1.

Table 1.2 Showing 100 random variates U distributed uniformly in the interval 3 to 5 obtained using program number 1.2 in Mathematica. u's used are also listed in this table.

i	u[i]	U[i]	i	u[i]	U[i]
1	0.8618	4.7237	51	0.7833	4.5667
2	0.4287	3.8574	52	0.7877	4.5755
3	0.8596	4.7192	53	0.7012	4.4024
4	0.0342	3.0685	54	0.5937	4.1875
5	0.7597	4.5193	55	0.6616	4.3232
6	0.4609	3.9217	56	0.5202	4.0403
7	0.4189	3.8378	57	0.9024	4.8049
8	0.4746	3.9491	58	0.0050	3.0100
9	0.3799	3.7598	59	0.2183	3.4365
10	0.1687	3.3374	60	0.2361	3.4722
11	0.1899	3.3799	61	0.4477	3.8954
12	0.0364	3.0729	62	0.7308	4.4615
13	0.3335	3.6669	63	0.3429	3.6857
14	0.4257	3.8514	64	0.2260	3.4520
15	0.3860	3.7721	65	0.1079	3.2158
16	0.2445	3.4890	66	0.4879	3.9758
17	0.5881	4.1763	67	0.1655	3.3310
18	0.2322	3.4644	68	0.1218	3.2437
19	0.6743	4.3486	69	0.3544	3.7088
20	0.9089	4.8179	70	0.3523	3.7046
21	0.8759	4.7518	71	0.3895	3.7789
22	0.3658	3.7316	72	0.0911	3.1823

23	0.2945	3.5891	73	0.9631	4.9262
24	0.6719	4.3438	74	0.5184	4.0368
25	0.9104	4.8208	75	0.6328	4.2656
26	0.9276	4.8552	76	0.8258	4.6516
27	0.0504	3.1008	77	0.4101	3.8201
28	0.3510	3.7019	78	0.6742	4.3484
29	0.1510	3.3021	79	0.2088	3.4176
30	0.1761	3.3523	80	0.8631	4.7262
31	0.3796	3.7593	81	0.2042	3.4084
32	0.5239	4.0479	82	0.3027	3.6054
33	0.3586	3.7172	83	0.0504	3.1007
34	0.4202	3.8404	84	0.8463	4.6925
35	0.5641	4.1283	85	0.4134	3.8267
36	0.1372	3.2743	86	0.2451	3.4903
37	0.0774	3.1548	87	0.5565	4.1130
38	0.8924	4.7849	88	0.0084	3.0169
39	0.9848	4.9696	89	0.3532	3.7064
40	0.4854	3.9707	90	0.7547	4.5094
41	0.3080	3.6159	91	0.1188	3.2376
42	0.4945	3.9890	92	0.5403	4.0806
43	0.9390	4.8781	93	0.5235	4.0471
44	0.5923	4.1846	94	0.8228	4.6456
45	0.1921	3.3842	95	0.5526	4.1052
46	0.3830	3.7660	96	0.8279	4.6557
47	0.6754	4.3508	97	0.6770	4.3540
48	0.2944	3.5888	98	0.7266	4.4531
49	0.9737	4.9474	99	0.4398	3.8796
50	0.4501	3.9001	100	0.2243	3.4485

1.3 Generating random variates by acceptance-rejection method

See Figure 1.4. Suppose, we wish to generate random variates x corresponding to a given probability density function $p(x)$ defined for the interval $x = a$ to b. Suppose, non-zero values of p span from 0 to c.

In acceptance-rejection method, 2 sets of uniform random numbers u_1's and u's in the interval 0 to 1 are first generated. Thereafter, linear transformations are carried out using $U_1 = a + (b-a)u_1$ and $U = 0 + (c-0)u$ known from sampling by inverse transform method (see equation (1.4)) to get 2 sets of uniform random numbers U_1's and U's in the interval a to b and 0 to c respectively. The next steps are to calculate $p(x = U_1)$ for each value of U_1 and if $U < p(x = U_1)$, that U_1 is accepted as a valid random variate for the probability density function $p(x)$; otherwise that value of U_1 is rejected.

Validity of the method narrated above can be understood as follows. Since U_1 is a uniform random variable in the interval a to b, probability that U_1 falls in the interval x to $x+dx$ is proportional to the width of the interval which is dx. Again, since U is a uniform random variable in the interval 0 to c, probability that U falls below $p(x = U_1)$ is proportional to the interval $p(x = U_1) -0 = p(x = U_1)$. Thus the probability that U_1 is within x to $x+dx$ and U is below $p(x = U_1)$ is (proportional to) the product of $p(x = U_1)$ and dx which is (proportional to) $p(x)dx$.

We now generate random variates corresponding to linear probability density function $p(x) = 2x/\pi^2$ in the interval $0 < x < \pi$ using acceptance-rejection sampling using program number 1.3. Part A of the program provides a table, see Table 1.3, containing acceptance-rejection sampling. Part B of the program provides a table, see Table 1.4, containing automated collection of the accepted values of the variate eta called etaa. Figure 1.5 reveals that indeed the 52 accepted random variates etaa are nearly linearly distributed in the interval 0 to π. For $0 < x < \pi$, we have $0 < p < 0.65$.

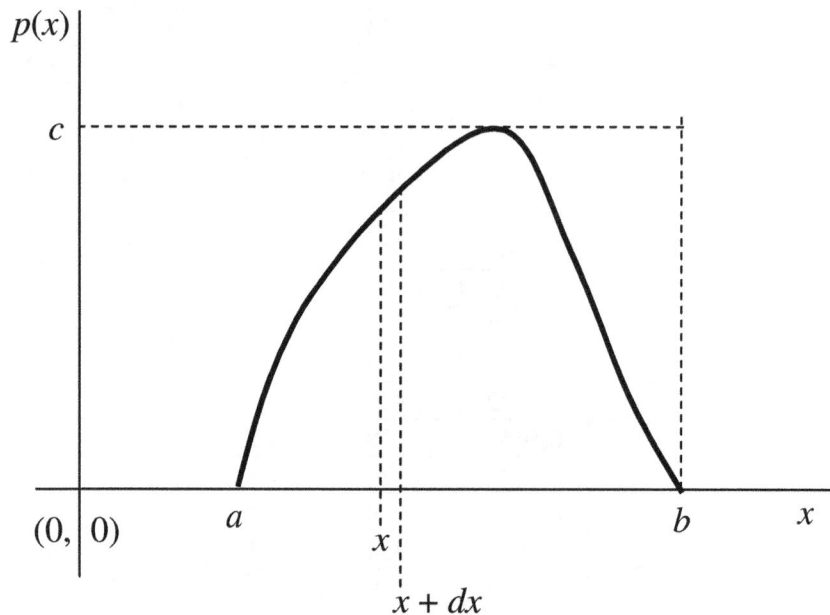

Figure 1.4 Figure to help explain acceptance-rejection method of obtaining random variates for a given probability density function $p(x)$.

Program number 1.3
part A

```
N1=100;

n=654321;

SeedRandom[n];

Table[{i=i+1,u1[i]=RandomReal[],u[i]=RandomReal[],

U1[i]=0+(Pi-0)*u1[i],U[i]=0+(0.65-0)*u[i],

If[U[i]<2*U1[i]/(Pi^2),eta[i]=U1[i],Null]},{i,0,N1-1,1}];

TableForm[%,TableSpacing->{2,2},

TableHeadings->{None,{"i","u1[i]","u[i]",

"U1[i]","U[i]","eta[i]"}}]
```

part B

```
etaa={};

i=0;

While[i<=N1,If[U[i]<2*U1[i]/(Pi^2),

AppendTo[etaa,U1[i]]];i=i+1]

n1=Length[etaa]


Table[{i=i+1,etaa[[i]]},{i,0,n1-1,1}];

TableForm[%,TableSpacing->{2,2},

TableHeadings->{None,{"i","etaa[[i]]" }}]
```

ListPlot[Table[{etaa[[i]],1},{i,0,n1-1,1}],Frame->True,

FrameLabel->{"etaa [[i]]","1"}]

Table 1.3 Showing acceptance-rejection sampling to get random variates eta obtained using part A of program number 1.3.

i	u1[i]	u[i]	U1[i]	U[i]	eta[i]
1	0.8618	0.4287	2.7075	0.2787	2.7075
2	0.8596	0.0342	2.7005	0.0223	2.7005
3	0.7597	0.4609	2.3865	0.2996	2.3865
4	0.4189	0.4746	1.3160	0.3085	Null
5	0.3799	0.1687	1.1935	0.1097	1.1935
6	0.1899	0.0364	0.5967	0.0237	0.5967
7	0.3335	0.4257	1.0476	0.2767	Null
8	0.3860	0.2445	1.2128	0.1589	1.2128
9	0.5881	0.2322	1.8477	0.1509	1.8477
10	0.6743	0.9089	2.1184	0.5908	Null
11	0.8759	0.3658	2.7517	0.2378	2.7517
12	0.2945	0.6719	0.9253	0.4367	Null
13	0.9104	0.9276	2.8601	0.6029	Null
14	0.0504	0.3510	0.1583	0.2281	Null
15	0.1510	0.1761	0.4745	0.1145	Null
16	0.3796	0.5239	1.1926	0.3406	Null
17	0.3586	0.4202	1.1266	0.2731	Null
18	0.5641	0.1372	1.7723	0.0892	1.7723
19	0.0774	0.8924	0.2431	0.5801	Null
20	0.9848	0.4854	3.0938	0.3155	3.0938
21	0.3080	0.4945	0.9675	0.3214	Null
22	0.9390	0.5923	2.9501	0.3850	2.9501
23	0.1921	0.3830	0.6035	0.2490	Null
24	0.6754	0.2944	2.1218	0.1914	2.1218
25	0.9737	0.4501	3.0590	0.2925	3.0590
26	0.7833	0.7877	2.4609	0.5120	Null
27	0.7012	0.5937	2.2028	0.3859	2.2028
28	0.6616	0.5202	2.0785	0.3381	2.0785
29	0.9024	0.0050	2.8351	0.0032	2.8351
30	0.2183	0.2361	0.6857	0.1534	Null
31	0.4477	0.7308	1.4065	0.4750	Null
32	0.3429	0.2260	1.0771	0.1469	1.0771
33	0.1079	0.4879	0.3390	0.3171	Null
34	0.1655	0.1218	0.5199	0.0792	0.5199
35	0.3544	0.3523	1.1133	0.2290	Null
36	0.3895	0.0911	1.2236	0.0592	1.2236
37	0.9631	0.5184	3.0256	0.3370	3.0256
38	0.6328	0.8258	1.9880	0.5368	Null
39	0.4101	0.6742	1.2882	0.4382	Null
40	0.2088	0.8631	0.6560	0.5610	Null
41	0.2042	0.3027	0.6416	0.1968	Null
42	0.0504	0.8463	0.1582	0.5501	Null
43	0.4134	0.2451	1.2986	0.1593	1.2986

44	0.5565	0.0084	1.7483	0.0055	1.7483
45	0.3532	0.7547	1.1096	0.4906	Null
46	0.1188	0.5403	0.3732	0.3512	Null
47	0.5235	0.8228	1.6447	0.5348	Null
48	0.5526	0.8279	1.7361	0.5381	Null
49	0.6770	0.7266	2.1269	0.4723	Null
50	0.4398	0.2243	1.3816	0.1458	1.3816
51	0.9524	0.7628	2.9922	0.4958	2.9922
52	0.9227	0.7673	2.8988	0.4988	2.8988
53	0.7744	0.0388	2.4329	0.0253	2.4329
54	0.4234	0.3146	1.3302	0.2045	1.3302
55	0.2849	0.8526	0.8952	0.5542	Null
56	0.2712	0.4589	0.8521	0.2983	Null
57	0.5662	0.1270	1.7787	0.0826	1.7787
58	0.3261	0.1921	1.0244	0.1249	1.0244
59	0.9294	0.0971	2.9197	0.0631	2.9197
60	0.8678	0.0527	2.7264	0.0342	2.7264
61	0.6010	0.5101	1.8880	0.3315	1.8880
62	0.0491	0.8662	0.1543	0.5630	Null
63	0.8311	0.3699	2.6111	0.2404	2.6111
64	0.6055	0.7086	1.9021	0.4606	Null
65	0.0621	0.0293	0.1951	0.0190	0.1951
66	0.9600	0.7918	3.0158	0.5146	3.0158
67	0.2790	0.3078	0.8764	0.2000	Null
68	0.9336	0.7173	2.9329	0.4662	2.9329
69	0.9645	0.1720	3.0300	0.1118	3.0300
70	0.1165	0.6613	0.3660	0.4299	Null
71	0.9249	0.6311	2.9056	0.4102	2.9056
72	0.6276	0.1130	1.9716	0.0734	1.9716
73	0.1215	0.9735	0.3816	0.6328	Null
74	0.0148	0.2807	0.0465	0.1824	Null
75	0.1744	0.8147	0.5479	0.5296	Null
76	0.9503	0.8674	2.9854	0.5638	2.9854
77	0.3100	0.7364	0.9738	0.4787	Null
78	0.2646	0.1094	0.8312	0.0711	0.8312
79	0.1454	0.9416	0.4569	0.6121	Null
80	0.3930	0.0684	1.2345	0.0445	1.2345
81	0.6325	0.3698	1.9870	0.2404	1.9870
82	0.3261	0.3225	1.0246	0.2096	Null
83	0.7350	0.4885	2.3089	0.3175	2.3089
84	0.4781	0.0687	1.5020	0.0447	1.5020
85	0.7332	0.7276	2.3035	0.4730	Null
86	0.2226	0.7403	0.6995	0.4812	Null
87	0.6044	0.2942	1.8987	0.1912	1.8987
88	0.3337	0.8206	1.0484	0.5334	Null
89	0.6196	0.1319	1.9464	0.0858	1.9464
90	0.4761	0.1548	1.4957	0.1006	1.4957
91	0.1495	0.1481	0.4696	0.0962	Null
92	0.0263	0.9820	0.0826	0.6383	Null
93	0.6774	0.4351	2.1281	0.2828	2.1281
94	0.3336	0.4401	1.0481	0.2861	Null

95	0.3045	0.2773	0.9565	0.1803	0.9565
96	0.9500	0.5186	2.9844	0.3371	2.9844
97	0.7374	0.9651	2.3167	0.6273	Null
98	0.6385	0.9689	2.0060	0.6298	Null
99	0.5705	0.4434	1.7922	0.2882	1.7922
100	0.5668	0.8242	1.7808	0.5357	Null

Table 1.4 Showing automated collection of the 52 accepted values of random variate eta denoted by etaa obtained using part B of program number 1.3.

i	etaa[[i]]
1	2.7075
2	2.7005
3	2.3865
4	1.1935
5	0.5967
6	1.2128
7	1.8477
8	2.7517
9	1.7723
10	3.0938
11	2.9501
12	2.1218
13	3.0590
14	2.2028
15	2.0785
16	2.8351
17	1.0771
18	0.5199
19	1.2236
20	3.0256
21	1.2986
22	1.7483
23	1.3816
24	2.9922
25	2.8988
26	2.4329
27	1.3302
28	1.7787
29	1.0244
30	2.9197
31	2.7264
32	1.8880
33	2.6111
34	0.1951
35	3.0158
36	2.9329
37	3.0300
38	2.9056

39	1.9716
40	2.9854
41	0.8312
42	1.2345
43	1.9870
44	2.3089
45	1.5020
46	1.8987
47	1.9464
48	1.4957
49	2.1281
50	0.9565
51	2.9844
52	1.7922

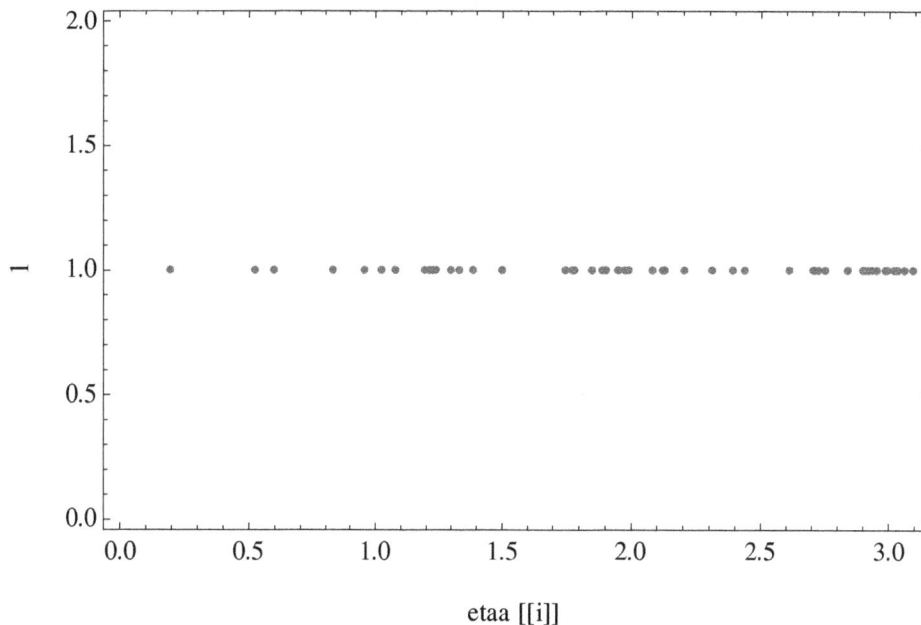

etaa [[i]]

Figure 1.5 Showing that the 52 accepted values of random variate eta are nearly linearly distributed in the interval 0 to π. There are more variates for larger values of etaa. Obtained using program number 1.3.

If probability density function p contains more than one variable, say x_1 and x_2, we first need to generate 3 sets of uniform random numbers u_1's, u_2's and u's in the interval 0 to 1. Thereafter, linear transformations are carried out using $U_1 = a + (b - a)u_1$, $U_2 = a + (b - a)u_2$ and $U = 0 + (c - 0)u$. We consider a set of 3 values: U_1, U_2, U and we calculate $p(x_1 = U_1, x_2 = U_2)$. If the condition $U < p(x_1 = U_1, x_2 = U_2)$ is satisfied, we accept *both* the values of U_1 and U_2 as valid random variate for the probability density function $p(x_1, x_2)$; see reference [1]. Extensions to probability density functions p containing more than two variables are evident. Practical uses of this are made in all of the remaining chapters of this book.

1.4 Variance reduction and importance sampling

Let us consider a definite integral

$$I = \int_a^b F(x)dx \qquad \text{--------(1.5)}$$

Let us re-write equation (1.5) as $I = \int_a^b \frac{F(x)}{p(x)} p(x)dx = \int_a^b f(x)p(x)dx \qquad \text{--------(1.6)}$

where $f(x) = F(x) / p(x)$.

Average value of any function of x, say $f(x)$, is given by $A_f = \int_a^b f(x)p(x)dx$ where $p(x)$ is a *normalized* probability density function. Thus according to equation (1.6), value I of the integral is average value of F / p.

We can choose any functional form for $p(x)$, but as proved below, choosing $p(x)$ as proportional to $F(x)$ ensures that the variance of $f(x) = F / p$ is small. This will ensure a better average value of F / p and hence a better value of the integral $I = \int_a^b F(x)dx$.

Variance of $f(x)$ is square of standard deviation. Variance of $f(x)$ is given by

$$V f(x) = A [(f - A_f)^2] \qquad \text{--------(1.7)}$$

where A stands for average value. Thus

$$V f(x) = A [f^2 - 2f A_f + A_f^2] = A(f^2) - 2A_f A_f + A_f^2 = A(f^2) - A_f^2$$

or, $V f(x) = \int_a^b f^2(x)p(x)dx - \left(\int_a^b f(x)p(x)dx\right)^2$

$$= \int_a^b \frac{F^2}{p^2} p\, dx - \left(\int_a^b \frac{F}{p} p dx\right)^2 \qquad = \int_a^b \frac{F^2}{p} dx - \left(\int_a^b F\, dx\right)^2$$

Thus the variance is

$$V f(x) = \int_a^b \frac{F^2}{p} dx - I^2 \qquad \text{--------(1.8)}$$

If $p(x)$ is taken as

$$p(x) = \frac{|F(x)|}{\int_a^b |F(x)|dx} \qquad \text{--------(1.9)}$$

the variance given by equation (1.8) becomes $V f(x) = \int_a^b \frac{F^2}{\dfrac{|F(x)|}{\int_a^b |F(x)|dx}} dx - I^2$

or, $V f(x) = \left(\int_a^b |F(x)|dx\right) \left(\int_a^b |F(x)|dx\right) - I^2 = \left(\int_a^b |F(x)|dx\right)^2 - I^2 \qquad \text{-------(1.10)}$

Equation (1.10) in conjunction with equation (1.5): $I = \int_a^b F(x)dx$ reveals that the variance vanishes if the integrand $F(x)$ does not change sign. If it does, the variance will be small if condition laid by equation (1.9) is met. Thus the probability density function $p(x)$ should be proportional to the integrand $|F(x)|$. This is the so called *importance sampling*.

13

Looking back at equation (1.6), we find that value of the integral is approximately determined by the value of the average $I = \dfrac{1}{N}\sum\limits_{i=1}^{N}\dfrac{F(x_i)}{p(x_i)}$. As recommended by importance sampling, we need to take $p(x) = C\,F(x)$ where C is a

constant; we need to normalize $p(x)$ first; $\int\limits_{a}^{b}p(x)dx = 1$ gives $\int\limits_{a}^{b}C\,F(x)dx = 1$ or, $C = \dfrac{1}{\int\limits_{a}^{b}F(x)dx}$. Thus $p(x) =$

$\dfrac{F(x)}{\int\limits_{a}^{b}F(x)dx}$. The sum $\dfrac{1}{N}\sum\limits_{i=1}^{N}\dfrac{F(x_i)}{p(x_i)}$ becomes $\dfrac{1}{N}\sum\limits_{i=1}^{N}\dfrac{F(x_i)}{\dfrac{F(x_i)}{\int\limits_{a}^{b}F(x)dx}} = \int\limits_{a}^{b}F(x)dx$. Thus the sum $\dfrac{1}{N}\sum\limits_{i=1}^{N}\dfrac{F(x_i)}{p(x_i)}$ equals

the integral if probability density function $p(x)$ is taken proportional to the integrand $\left|F(x)\right|$. Slight variation of $p(x)$

from proportionality with $\left|F(x)\right|$ will result in slight difference of $\dfrac{1}{N}\sum\limits_{i=1}^{N}\dfrac{F(x_i)}{p(x_i)}$ from actual value of the integral $I =$

$\int\limits_{a}^{b}F(x)dx$.

1.5 A 1D example: evaluation of definite integral using acceptance-rejection sampling

As to a 1D example, we now take up the integral

$$I = \int\limits_{a}^{b}F(x)dx = \int\limits_{0}^{2}e^{x}dx \qquad \text{-------(1.11)}$$

where $F(x) = e^{x}$. We re-write equation (1.11) as

$$I = \int\limits_{a}^{b}\dfrac{F(x)}{p(x)}p(x)dx \qquad \text{-------(1.12)}$$

which, as discussed in section 1.4, implies that average value of F/p is the value of the integral, i.e.

$$I = \dfrac{1}{N}\sum\limits_{i=1}^{N}\dfrac{F(x_i)}{p(x_i)} = \dfrac{1}{N}\sum\limits_{i=1}^{N}y_i \qquad \text{-------(1.13)}$$

where x_i's are random values of x in the interval $a < x < b$ obeying probability density function $p(x)$ and $y_i = \dfrac{F(x_i)}{p(x_i)}$.

We now take up a linear variation for $p(x)$ given by $p(x) = C\,x$ where C is a constant. Normalization of $p(x)$ requires $\int\limits_{0}^{2}p(x)dx = 1$ or, $C = 1/2$. As such normalized probability density function is $p(x) = x/2$. See Figure 1.6 in which we have plotted $F(x) = e^{x}$ along with the normalized linear probability density function $p(x) = x/2$. The linear probability density function $p(x) = x/2$, although *not* proportional to (the variation of) e^{x}, follows the function e^{x} to some extent, i.e. both are increasing functions of x (in the chosen interval $0 < x < 2$).

We now generate accepted values of random variate UA1[[i]] obeying the probability density function $p(x) = x/2$ in the interval $0 < x < 2$ using acceptance-rejection sampling, using program number 1.4. The n_1 accepted values are shown in

Table 1.5. Using equation (1.13): $I = \dfrac{1}{n_1}\sum\limits_{i=1}^{n_1}\dfrac{F(x_i)}{p(x_i)} = \dfrac{1}{n_1}\sum\limits_{i=1}^{n_1}y_i = \dfrac{1}{n_1}\sum\limits_{i=1}^{n_1}\dfrac{e^{x_i}}{x_i/2} = \dfrac{1}{n_1}\sum\limits_{i=1}^{n_1}\dfrac{e^{UA1[[i]]}}{UA1[[i]]/2}$ in program

number 1.4, we evaluate the integral with the result 6.34805 rather than the standard value 6.38906.

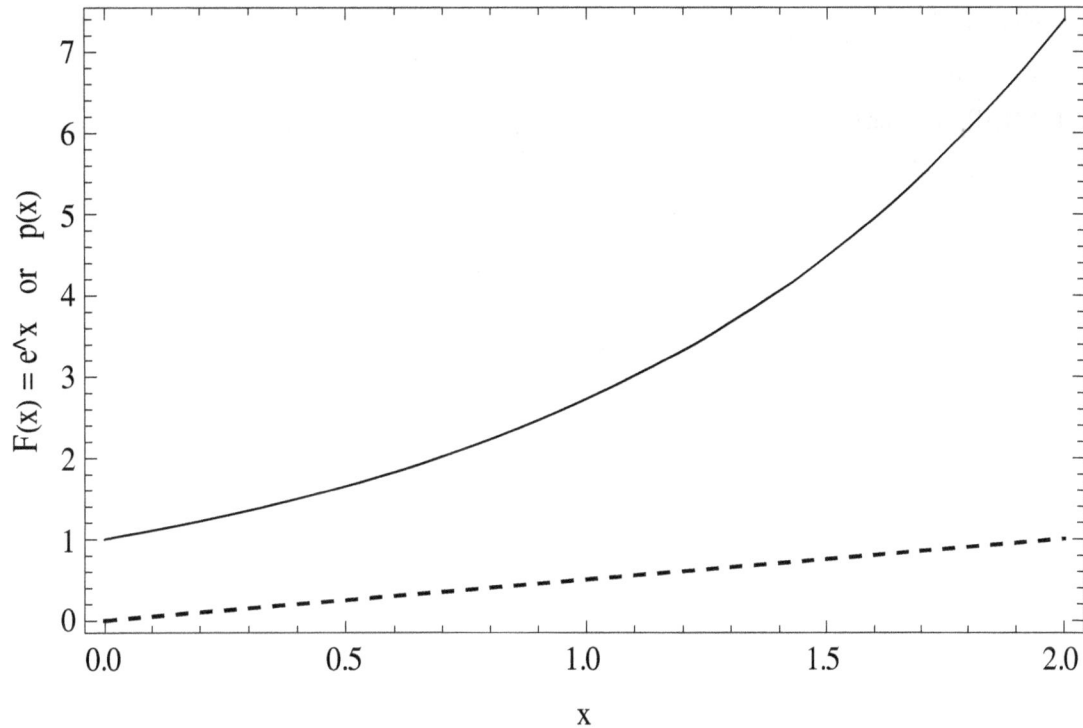

Figure 1.6 Showing $F(x) = e^x$ as undashed curve and $p(x) = x/2$ as dashed curve obtained using the command
Plot[{Exp[x],x/2},{x,0,2},Frame->True,
FrameLabel->{"x","F(x) = e^x   or   p(x)"},
PlotStyle->{{Black},{Dashed, Black}}]
in Mathematica.

Program number 1.4

N1=150;

a=0;

b=2;

c=1;

n=654321;

SeedRandom[n];

Table[{i=i+1,U1[i]=a+(b-a)*RandomReal[],U[i]=0+(c-0)*RandomReal[]},{i,0,N1-1,1}];

UA1={};

i=0;

While[i<=N1,If[U[i]<U1[i]/2,AppendTo[UA1,U1[i]]];i=i+1];

n1=Length[UA1]

Table[{i=i+1,UA1[[i]] },{i,0,n1-1,1}];

TableForm[%,TableSpacing->{2,2},TableHeadings->{None,{"i","UA1[[i]]" }}];

Table[{i=i+1,UA1[[i]],y[i]=(Exp[UA1[[i]]])/ (UA1[[i]]/2)},{i,0,n1-1,1}];

TableForm[%,TableSpacing->{2,2},

TableHeadings->{None,{"i","UA1[[i]]","y[i]"}}]

Integral=Sum[y[i]/n1,{i,1,n1}]

ListPlot[Table[{UA1[[i]],1},{i,0,n1-1,1}],

Frame->True,FrameLabel->{"UA1[[i]]","1"}]

That accepted values of random variate UA1[[i]] are distributed as per probability density function $p(x) = x/2$ in the interval 0 to 2 is evidenced by Figure 1.7 which displays values of UA1[[i]] against stagnant integer 1. We find that indeed there are more variates for larger values of UA1[[i]].

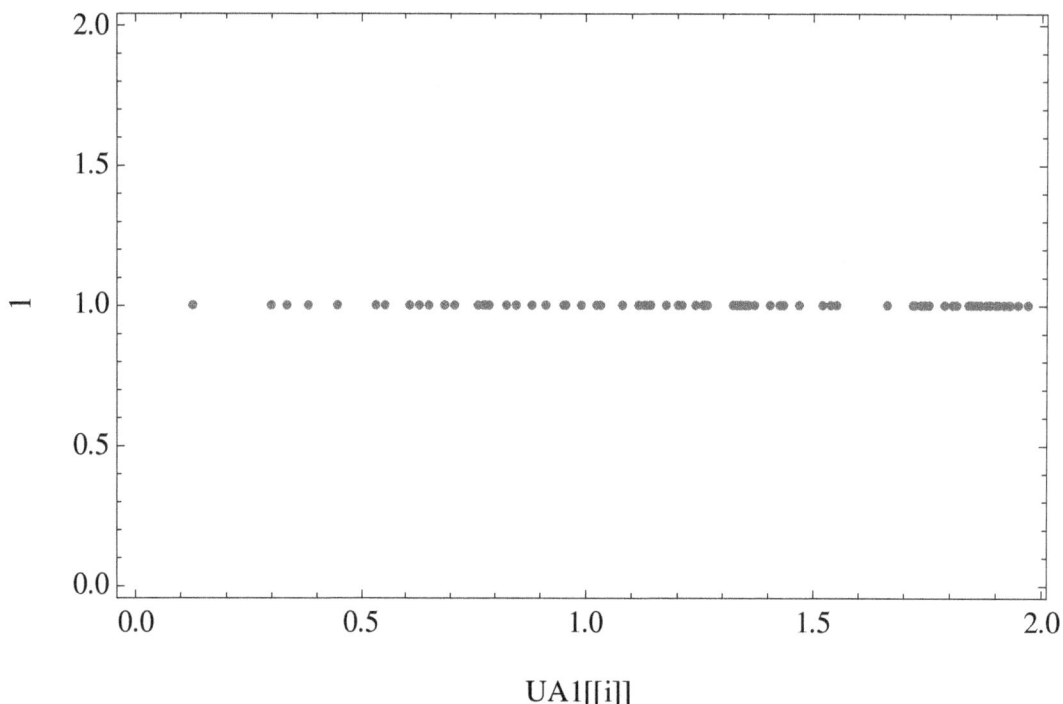

UA1[[i]]

Figure 1.7 Showing that values of random variate UA1[[i]] obtained using program number 1.4 are almost linearly distributed in the interval 0 to 2. There are more variates for larger values of UA1[[i]].

Table 1.5 Showing tabulated values of UA1[[i]] and y[i] = (Exp[UA1[[i]]])/(UA1[[i]]/2) obtained using program number 1.4

i	UA1[[i]]	y[i]	i	UA1[[i]]	y[i]
1	1.7237	6.5036	51	0.9522	5.4430
2	1.7192	6.4915	52	0.2990	9.0205
3	1.5193	6.0147	53	1.3548	5.7219
4	0.7598	5.6274	54	0.6089	6.0384
5	0.3799	7.6977	55	1.9000	7.0376
6	0.7721	5.6063	56	1.1410	5.4862

7	1.1763	5.5128	57	1.4322	5.8482
8	1.7518	6.5817	58	1.8862	6.9920
9	1.1283	5.4780	59	1.5364	6.0504
10	1.9696	7.2783	60	1.0791	5.4527
11	1.8781	6.9655	61	0.9103	5.4599
12	1.3508	5.7159	62	0.5527	6.2888
13	1.9474	7.1999	63	1.7449	6.5624
14	1.4024	5.7971	64	1.0243	5.4381
15	1.3232	5.6763	65	1.1769	5.5133
16	1.8049	6.7364	66	0.7869	5.5829
17	0.6857	5.7902	67	1.3311	5.6873
18	0.3310	8.4130	68	1.8652	6.9238
19	0.7088	5.7324	69	1.3411	5.7017
20	0.7789	5.5952	70	1.8578	6.9001
21	1.9262	7.1262	71	0.9874	5.4370
22	0.8267	5.5298	72	0.4450	7.0136
23	1.1130	5.4690	73	1.7206	6.4952
24	0.8796	5.4796	74	1.3704	5.7456
25	1.9049	7.0541	75	1.8395	6.8426
26	1.8455	6.8612	76	1.7345	6.5335
27	1.5488	6.0767	77	1.3460	5.7088
28	0.8468	5.5082	78	0.6315	5.9553
29	1.1323	5.4805	79	1.2623	5.5985
30	0.6521	5.8873	80	1.0337	5.4396
31	1.8587	6.9032	81	1.8156	6.7688
32	1.7357	6.5366	82	1.4258	5.8369
33	1.2019	5.5353	83	1.7864	6.6816
34	1.6623	6.3422	84	1.9035	7.0493
35	0.1242	18.2282			
36	1.9199	7.1050			
37	1.8672	6.9302			
38	1.9289	7.1357			
39	1.8498	6.8747			
40	1.2552	5.5904			
41	1.9006	7.0398			
42	0.5292	6.4158			
43	0.7859	5.5843			
44	1.2650	5.6017			
45	0.6523	5.8869			
46	1.4699	5.9172			
47	0.9562	5.4419			
48	1.4665	5.9107			
49	1.2087	5.5417			
50	1.2391	5.5726			

A key take-away from this section is that probability density function $p(x)$ does not have to be proportional to $F(x)$ in practice. Satisfactory value of integral can be obtained if $p(x)$ follows $F(x)$; i.e. if $F(x)$ is an increasing function of x, $p(x)$ also needs to be an increasing function of x (being same in order of magnitude).

1.6 The case with higher-dimension: evaluation of definite integral and acceptance-rejection sampling

We now turn to the case if the integrand contains more than one variable. Let the definite integral be $I = \int_a^b \int_a^b F(x_1, x_2) \; dx_1 \; dx_2$. We need to write it as

$$I = \int_a^b \int_a^b \frac{F(x_1, x_2)}{p(x_1, x_2)} p(x_1, x_2) \; dx_1 \; dx_2$$

where, **first** of all, the probability density function p has to be *normalized*. **Secondly**, p must be chosen such that p follows F if not be proportional to F. That is, if F is an increasing function of say x_1, p has to be an increasing function of x_1 also. If F is a decreasing function of say x_2, p has to be a decreasing function of x_2 also.

Thirdly, as already described at the end of section 1.3, random variates for x_1 and x_2 are to be obtained as follows (repeat):

If probability density function p contains more than one variable, say x_1 and x_2, we first need to generate 3 sets of uniform random numbers u_1's, u_2's and u's in the interval 0 to 1. Thereafter, linear transformations are carried out using $U_1 = a + (b - a)u_1$, $U_2 = a + (b - a)u_2$ and $U = 0 + (c - 0)u$. We consider a set of 3 values: U_1, U_2, U and we calculate $p(x_1 = U_1, x_2 = U_2)$. If the condition $U < p(x_1 = U_1, x_2 = U_2)$ is satisfied, we accept *both* the values of U_1 and U_2 as valid random variate for the probability density function $p(x_1, x_2)$; see reference [1]. Accepted values of $U1$ and $U2$ are denoted by UA1 and UA2 respectively. Extensions to probability density functions p containing more than two variables are evident. Practical uses of this are made in all of the remaining chapters of this book.

Thereafter, the integral is evaluated as follows.

$$I = \frac{1}{n_1 \; n_2} \sum_{i=1}^{n1} \sum_{j=1}^{n2} \frac{F(x_{1i}, x_{2j})}{p(x_{1i}, x_{2j})}$$

or, $$I = \frac{1}{n_1 \; n_2} \sum_{i=1}^{n1} \sum_{j=1}^{n2} \frac{F(UA1[[i]], \; UA2[[j]])}{p(UA1[[i]], \; UA2[[j]])}$$

where n_1 and n_2 are number of accepted variates for x_1 and x_2 respectively. Extensions to 3 or higher dimensional integrations are evident. Practical uses are demonstrated in all of the remaining chapters of this book.

As to the integrand $F = (x1/(x2+x3))^2$ for example, we can choose the probability density function p as $(x1)*(1/x2)*(1/x3)$. There are several virtues of this probability density function as required. **Firstly**, this probability density function can be normalized without running into the problem of evaluating another multi-dimensional definite integral. **Secondly**, both F and p are increasing functions of x_1, and decreasing functions of x_2 and x_3.

If random variates UA1[i], UA2[j] and UA3[k] are obtained using acceptance-rejection sampling as applied to multi-variable probability density function, we can evaluate the integration as

$$I = \frac{1}{n_1 \; n_2 \; n_3} \sum_{i=1}^{n1} \sum_{j=1}^{n2} \sum_{k=1}^{n3} \frac{F(UA1[[i]], \; UA2[[j]], \; UA3[[k]])}{p(UA1[[i]], \; UA2[[j]], \; UA3[[k]])}$$

or, $$I = \frac{1}{n_1 \; n_2 \; n_3} \sum_{i=1}^{n1} \sum_{j=1}^{n2} \sum_{k=1}^{n3} \frac{(UA1[[i]]/(UA2[[j]] + UA3[[k]]))^2}{norc*(UA1[[i]]*(1/UA2[[j]])*(1/UA3[[k]]))}$$

Here n_1, n_2 and n_3 are number of accepted variates for x_1, x_2 and x_3 respectively.

If a to b are limits of the integration, maximum value of probability density function p is $c = norc*(b*(1/a)*(1/a))$ where *norc* is normalization constant. This information is required for the acceptance-rejection sampling.

Chapter II

Evaluation of Two-dimensional Definite Integrals

This chapter deals with Monte Carlo evaluation of 3 two-dimensional definite integrals. In each case, a suitable multi-variable probability density function is chosen and acceptance-rejection sampling is used to obtain values of 2 sets of random variates corresponding to the 2 random variables in the probability density function. The integrals are evaluated using the 2 sets of random variates. Programs written in Mathematica have been used in the sampling as well as in evaluating the integrals. Uses of different parts of the programs have been narrated.

2.1 The 2D definite integrals dealt with in this chapter
This chapter deals with the following 3 definite integrals:

$$I_1 = \int_a^b \int_a^b (x_1 + x_2)^y \, dx_1 \, dx_2 \qquad \text{------(2.1)}$$

$$I_2 = \int_a^b \int_a^b (x_1 + x_2)^{-y} \, dx_1 \, dx_2 \qquad \text{------(2.2)}$$

and
$$I_3 = \int_{x1=a}^b \int_{x2=g}^h (x_1 + x_2)^{-y} \, dx_1 \, dx_2 \qquad \text{------(2.3)}$$

where y is positive constant such as 0.5, 1, 2 etc.; limits a and b are also constants such as 1, 5, 11, 21, 31 etc.

2.2 Evaluation of the integral I_1

We first deal with the definite integral $I_1 = \int_a^b \int_a^b (x_1 + x_2)^y \, dx_1 \, dx_2$ where y is positive constant. We first choose a probability density function

$$p(x_1, x_2) = norc \ (x_1) \ (x_2) \qquad \text{-------(2.4)}$$

where $norc$ is normalization constant. Both $F(x_1, x_2) = (x_1 + x_2)^y$ and $p(x_1, x_2) = norc \ (x_1) \ (x_2)$ are increasing functions of both x_1 and x_2. The probability density function can be normalized without running into the problem of evaluating another multi-dimensional definite integral. Maximum value of the probability density function is $c = norc \ (b)(b)$.

Program number 2.1
part A

y=0.5

N1=150;

a=1;

b=11;

norc=1/(NIntegrate[(x1)*(x2),{x1,a,b},{x2,a,b}])

c=norc*(b)*(b)

n=654321;

SeedRandom[n];

Table[{i=i+1,

U1[i]=a+(b-a)*RandomReal[],

U2[i]=a+(b-a)*RandomReal[],

u[i]=RandomReal[];U[i]=0+(c-0)*u[i]},{i,0,N1-1,1}];

TableForm[%,TableSpacing->{2,2},

TableHeadings->{None,{"i","U1[i]","U2[i]","U[i]"}}]

part B

UA1={};

i=0;

While[i<=N1,If[U[i]<norc*(U1[i])*(U2[i]),

AppendTo[UA1,U1[i]]];i=i+1];

n1=Length[UA1]

UA2={};

i=0;

While[i<=N1,If[U[i]<norc*(U1[i])*(U2[i]),

AppendTo[UA2,U2[i]]];i=i+1];

n2=Length[UA2]

Table[{i=i+1,UA1[[i]],UA2[[i]]},{i,0,n1-1,1}];

TableForm[%,TableSpacing->{2,2},

TableHeadings->{None,{"i","UA1[[i]]","UA2[[i]]"}}]

part C

Imc=(1/(n1*n2))*Sum[(((UA1[[i]]+UA2[[j]])^y)/

(norc*(UA1[[i]])*(UA2[[j]])),{i,1,n1},{j,1,n2}]

Ini=NIntegrate[(x1+x2)^y,{x1,a,b},{x2,a,b}]

error1=(Imc-Ini)*100/Ini

To evaluate the 2D definite integral I_1, we have written program number 2.1. In part A of program number 2.1, $U1$ and $U2$ are 2 sets of $N1$ uniform random numbers for x_1 and x_2 respectively in the interval a to b. u and U are 2 sets of $N1$ uniform random numbers in interval 0 to 1 and 0 to c respectively. Table 2.1 shows these random numbers $U1$, $U2$, U for $N1 = 150$.

Table 2.1 Showing values of random numbers obtained using part A of program number 2.1 for $N1 = 150$. The program performs acceptance-rejection sampling using these random numbers.

i	$U1[i]$	$U2[i]$	$U[i]$
1	9.6183	5.2872	0.0289
2	1.3424	8.5966	0.0155
3	5.1891	5.7457	0.0128
4	2.6870	2.8994	0.0012
5	4.3347	5.2569	0.0130
6	3.4448	6.8815	0.0078
7	7.7431	10.0893	0.0294
8	4.6582	3.9453	0.0226
9	10.1039	10.2759	0.0017
10	4.5095	2.5104	0.0059

...	...	...	...
141	8.1009	7.4208	0.0264
142	7.3137	3.0609	0.0173
143	7.0132	2.0203	0.0017
144	5.3940	10.8578	0.0114
145	6.6114	9.8325	0.0089
146	6.6517	3.8128	0.0177
147	6.9845	10.6348	0.0109
148	1.2793	3.5885	0.0099
149	10.1107	4.2495	0.0205
150	3.6045	3.2336	0.0020

Part B of program number 2.1 uses the random numbers of Table 2.1 according to acceptance-rejection sampling method and produces 2 sets of $n_1 = n_2 = 52$ accepted random variates UA1 and UA2 for x_1 and x_2 respectively shown in Table 2.2 for $N1 = 150$. Acceptance-rejection sampling technique uses the condition $U < norc\ (U1)(U2)$. If the condition is satisfied, both $U1$ and $U2$ are accepted as random variates which we call $UA1[[i]]$ and $UA2[[i]]$ respectively for further calculations.

Table 2.2 Showing accepted values of random variates obtained using part B of program number 2.1. We have 52 accepted values out of 150 for both $U1$ and $U2$.

i	UA1[[i]]	UA2[[i]]
1	2.6870	2.8994
2	10.1039	10.2759
3	5.2020	6.6414
4	10.3903	6.9229
5	4.8300	7.7540
6	5.4771	8.3075
7	4.5231	4.8947
8	3.0881	9.6310
9	7.7702	8.2656
10	3.8494	9.5263
11	5.5895	6.6617
12	1.9707	9.6783
13	7.0096	6.1007
14	9.6619	9.3113
15	7.0547	8.0861
16	8.1727	10.6447
17	7.3108	7.2759
18	2.2148	10.7352
19	10.5029	9.6737
20	8.3643	3.6459
21	5.8852	5.7810
22	8.3323	8.2764
23	8.4025	7.0437
24	2.4949	2.4806
25	10.8203	7.7740
26	8.3743	10.6507
27	10.6886	6.7049
28	8.6822	6.9195
29	9.7804	6.3954
30	3.5408	7.0587

31	10.8582	4.9343
32	3.1282	10.8253
33	10.1837	7.6553
34	9.1117	10.2888
35	7.1196	9.6727
36	6.9957	10.5173
37	5.1004	9.5586
38	7.9099	5.1186
39	7.6092	3.5696
40	9.2763	9.7926
41	8.8762	5.6234
42	5.7033	8.8761
43	10.7184	1.2935
44	8.8484	10.4154
45	8.9531	9.9953
46	5.5710	6.5660
47	2.9332	9.2583
48	7.0132	2.0203
49	5.3940	10.8578
50	6.6114	9.8325
51	6.9845	10.6348
52	3.6045	3.2336

Using part C of program number 2.1, we evaluate the integral I_1 using the following equation

$$I_1 = \frac{1}{n_1 \, n_2} \sum_{i=1}^{n_1} \sum_{j=1}^{n_2} \frac{F(x_{1i}, x_{2j})}{p(x_{1i}, x_{2j})} = \frac{1}{n_1 \, n_2} \sum_{i=1}^{n_1} \sum_{j=1}^{n_2} \frac{(x_{1i} + x_{2j})^y}{norc \ (x_{1i})(x_{2j})}$$

as
$$I_1 = \frac{1}{n_1 \, n_2} \sum_{i=1}^{n_1} \sum_{j=1}^{n_2} \frac{(\ UA1[[i]] + UA2[[j]] \ )^y}{norc \ (\ UA1[[i]] \ ) \ (\ UA2[[j]] \ )}$$
--------(2.5)

Results are in Table 2.3 as Monte Carlo result I_{mc}.

In part C of program number 2.1, we have also calculated the integral as I_{ni} using NIntegrate command. Results are in Table 2.3 as I_{ni}. We have compared I_{ni} with I_{mc} for various values of parameters and of $N1$. The difference is denoted as error1 which is % error shown in Table 2.3.

Table 2.3 Results of part C of program number 2.1 for various values of parameters. We are dealing with
$$I_1 = \int_a^b \int_a^b (x_1 + x_2)^y \, dx_1 \, dx_2$$

i	y	a	b	$N1$	$n1$	$n1^2$	Monte Carlo result I_{mc}	result obtained using NIntegrate command I_{ni}	% error
1	2	1	5	50	20	400	617.42	618.67	-0.20
2	2	1	11	50	18	324	16333.40	16066.70	1.66
3	2	1	21	50	17	289	228952.00	220267.00	3.94
4	2	1	31	50	15	225	1.09*10^6	1.06*10^6	3.64
5	2	1	41	50	14	196	3.42*10^6	3.25*10^6	5.12
6	2	1	51	50	14	196	8.24*10^6	7.80*10^6	5.55
7	2	1	71	50	13	169	3.03*10^7	2.94*10^7	3.22
8	2	1	101	50	13	169	1.25*10^8	1.21*10^8	3.61

9	1	1	11	50	18	324	1272.23	1200.00	6.02
10	1	1	21	50	17	289	9794.60	8800.00	11.30
11	1	1	21	75	28	784	9270.29	8800.00	5.34
12	1	1	21	100	34	1156	8900.27	8800.00	1.14
13	1	1	31	100	32	1024	29151.60	28800.00	1.22
14	1	1	41	100	31	961	68893.80	67200.00	2.52
15	1	1	51	100	31	961	134093.00	130000.00	3.15
16	1	1	71	100	29	841	321161.00	352800.00	-8.97
17	1	1	71	150	43	1849	336182.00	352800.00	-4.71
18	1	1	101	150	43	1849	976189.00	1.02*10^6	-4.30
19	0.5	1	11	150	52	2704	352.15	340.84	3.32
20	0.5	1	21	150	48	2304	1867.55	1839.30	1.54
21	0.5	1	31	150	46	2116	5099.21	4983.49	2.32
22	0.5	1	41	150	45	2025	10548.30	10141.30	4.01
23	0.5	1	51	150	45	2025	18486.50	17622.20	4.90
24	0.5	1	71	150	43	1849	37948.60	40616.50	-6.57
25	0.5	1	101	150	43	1849	92720.50	98610.40	-5.97

Table 2.3 displays a survey of Monte Carlo evaluation of the 2D definite integration I_1 for different values of various parameters. We note that, while using Monte Carlo method, % error is generally below 10%. If the error is more than 10%, the error can usually be lowered by raising number of terms $n1^2$ in the summation in equation (2.5) by raising value of $N1$.

2.3 Evaluation of the integral I_2

We now deal with the definite integral $I_2 = \int_a^b \int_a^b (x_1 + x_2)^{-y} dx_1 \, dx_2$ where y is positive constant. We first choose a probability density function

$$p(x_1, x_2) = norc \ (1/x_1) \ (1/x_2) \qquad\qquad \text{-------(2.6)}$$

where $norc$ is normalization constant. Both $F(x_1, x_2) = (x_1 + x_2)^{-y}$ and $p(x_1, x_2) = norc \ (1/x_1) \ (1/x_2)$ are decreasing functions of both x_1 and x_2. The probability density function can be normalized without running into the problem of evaluating another multi-dimensional definite integral. Maximum value of the probability density function is $c = norc \ (1/a)(1/a)$.

Program number 2.2
part A

```
y=2
N1=200;
a=1;
b=11;
norc=1/(NIntegrate[(1/x1)*(1/x2),{x1,a,b},{x2,a,b}])
c=norc*(1/a)*(1/a)
n=654321;
SeedRandom[n];
Table[{i=i+1,
U1[i]=a+(b-a)*RandomReal[],
```

```
U2[i]=a+(b-a)*RandomReal[],
u[i]=RandomReal[];U[i]=0+(c-0)*u[i]},{i,0,N1-1,1}];
TableForm[%,TableSpacing->{2,2},
TableHeadings->{None,{"i","U1[i]","U2[i]","U[i]"}}]
```

part B

```
UA1={};
i=0;
While[i<=N1,If[U[i]<norc*(1/(U1[i]*U2[i])),
AppendTo[UA1,U1[i]]];i=i+1];
n1=Length[UA1]


UA2={};
i=0;
While[i<=N1,If[U[i]<norc*(1/(U1[i]*U2[i])),
AppendTo[UA2,U2[i]]];i=i+1];
n2=Length[UA2]


Table[{i=i+1,UA1[[i]],UA2[[i]]},{i,0,n1-1,1}];
TableForm[%,TableSpacing->{2,2},
TableHeadings->{None,{"i","UA1[[i]]","UA2[[i]]" }}]
```

part C

```
Imc=(1/(n1*n2))*Sum[(((UA1[[i]]+UA2[[j]])^(-y))/
(norc*(1/(UA1[[i]]*UA2[[j]])))),{i,1,n1},{j,1,n2}]
Ini=NIntegrate[(x1+x2)^(-y),{x1,a,b},{x2,a,b}]
error1=(Imc-Ini)*100/Ini
```

To evaluate the 2D definite integral I_2, we have written program number 2.2. In part A of program number 2.2, $U1$ and $U2$ are 2 sets of $N1$ uniform random numbers for x_1 and x_2 respectively in the interval a to b. u and U are 2 sets of $N1$ uniform random numbers in interval 0 to 1 and 0 to c respectively. Table 2.4 shows these random numbers $U1$, $U2$, U for $N1 = 200$.

Table 2.4 Showing values of random numbers obtained using part A of program number 2.2 for $N1 = 200$. The program performs acceptance-rejection sampling using these random numbers.

i	$U1[i]$	$U2[i]$	$U[i]$
1	9.6183	5.2872	0.1495
2	1.3424	8.5966	0.0802
3	5.1891	5.7457	0.0661
4	2.6870	2.8994	0.0063

5	4.3347	5.2569	0.0671
6	3.4448	6.8815	0.0404
7	7.7431	10.0893	0.1523
8	4.6582	3.9453	0.1169
9	10.1039	10.2759	0.0088
10	4.5095	2.5104	0.0306
...	...	...	...
191	1.2308	1.4819	0.1340
192	9.9166	5.9075	0.0557
193	9.9450	9.3613	0.1069
194	8.0495	9.9891	0.1063
195	4.9758	10.8989	0.1086
196	2.3146	2.2417	0.0498
197	4.0121	2.3987	0.0007
198	2.9337	1.4852	0.1529
199	3.5455	10.6181	0.1532
200	9.4406	8.6826	0.1529

Part B of program number 2.2 uses the random numbers of Table 2.4 according to acceptance-rejection sampling method and produces 2 sets of $n_1 = n_2 = 9$ accepted random variates $UA1$ and $UA2$ for x_1 and x_2 respectively shown in Table 2.5 for $N1 = 200$. Acceptance-rejection sampling technique uses the condition $U < norc\ (1/U1)(1/U2)$. If the condition is satisfied, both $U1$ and $U2$ are accepted as random variates which we call $UA1[[i]]$ and $UA2[[i]]$ respectively for further calculations.

Table 2.5 Showing accepted values of random variates obtained using part B of program number 2.2. We have 9 accepted values out of 200 for both $U1$ and $U2$.

i	UA1[[i]]	UA2[[i]]
1	2.6870	2.8994
2	1.0499	3.1826
3	2.2148	10.7352
4	2.4949	2.4806
5	3.5408	7.0587
6	7.0132	2.0203
7	3.6045	3.2336
8	5.2244	4.8979
9	4.0121	2.3987

Using part C of program number 2.2, we evaluate the integral I_2 using the following equation

$$I_2 = \frac{1}{n_1\,n_2} \sum_{i=1}^{n_1} \sum_{j=1}^{n_2} \frac{F(x_{1i}, x_{2j})}{p(x_{1i}, x_{2j})} = \frac{1}{n_1\,n_2} \sum_{i=1}^{n_1} \sum_{j=1}^{n_2} \frac{(x_{1i} + x_{2j})^{-y}}{norc\ (1/x_{1i})(1/x_{2j})}$$

as $$I_2 = \frac{1}{n_1\,n_2} \sum_{i=1}^{n_1} \sum_{j=1}^{n_2} \frac{(\ UA1[[i]] + UA2[[j]]\ )^{-y}}{norc\ (\ 1/UA1[[i]]\ )\ (\ 1/UA2[[j]]\ )} \qquad \text{-------(2.7)}$$

Results are in Table 2.6 as Monte Carlo results I_{mc}.

In part C of program number 2.2, we have also calculated the integral as I_{ni} using NIntegrate command. Results are in Table 2.6 as I_{ni}. We have compared I_{ni} with I_{mc} for various values of parameters and of $N1$. The difference is denoted as error1 which is % error shown in Table 2.6.

Table 2.6 Results of part C of program number 2.2 for various values of parameters. We are dealing with

$$I_2 = \int_a^b \int_a^b (x_1 + x_2)^{-y} dx_1 \, dx_2$$

i	y	a	b	$N1$	$n1$	$n1^2$	Monte Carlo result I_{mc}	result obtained using NIntegrate command I_{ni}	% error
1	2	1	5	50	10	100	0.59	0.59	-0.24
2	2	1	11	200	9	81	1.28	1.19	7.84
3	2	1	21	400	10	100	1.74	1.75	-0.39
4	2	1	31	600	9	81	2.04	2.11	-3.33
5	2	1	41	1200	9	81	2.48	2.38	4.56
6	2	1	51	1500	11	121	2.76	2.58	6.69
7	2	1	71	2500	9	81	3.20	2.90	10.29
8	2	1	71	7000	21	441	3.01	2.90	3.55
9	2	1	101	5000	11	121	3.65	3.25	12.43
10	2	1	101	10000	23	529	3.35	3.25	3.07
11	1	1	11	200	9	81	9.83	9.75	0.82
12	1	1	21	500	12	144	22.83	22.36	2.11
13	1	1	31	1000	11	121	38.03	35.46	7.25
14	1	1	51	1800	13	169	39.87	62.20	-35.90
15	1	1	51	5000	30	900	53.31	62.20	-14.30
16	1	1	51	10000	56	3136	57.54	62.20	-7.50
17	0.5	1	11	200	9	81	28.85	30.51	-5.41
18	0.5	1	31	1000	11	121	195.97	171.97	13.95
19	0.5	1	31	1200	17	289	186.46	171.97	8.43
20	0.5	1	31	1300	18	324	173.49	171.97	0.88
21	0.5	1	51	10000	56	3136	328.79	377.36	-12.87
22	0.5	1	51	13000	83	6889	351.37	377.36	-6.89
23	0.5	1	51	15000	95	9025	366.70	377.36	-2.83

Table 2.6 displays a survey of Monte Carlo evaluation of the 2D definite integration I_2 for different values of various parameters. We note that, while using Monte Carlo method, % error is generally below 10%. If the error is more than 10%, the error can usually be lowered by raising number of terms $n1^2$ in the summation in equation (2.7) by raising value of $N1$.

2.4 Evaluation of the integral I_3

We now evaluate the multi-dimensional definite integral

$$I_3 = \int_{x_1=a}^{b} \int_{x_2=g}^{h} (x_1 + x_2)^{-y} dx_1 \, dx_2 \qquad \text{------(2.8)}$$

using acceptance-rejection sampling. Here a, b, g, h are various constants, $y = 0.5, 1, 1.5$ and 2. We have written program number 2.3 in Mathematica using symbolic computation.

Program number 2.3
part A

y=2

N1=N2=200;

a=1;

b=11;

```
g=1;
h=21;
norc=1/(NIntegrate[(1/x1)*(1/x2),{x1,a,b},{x2,g,h}])
c=norc*(1/a)*(1/g)
n=654321;
SeedRandom[n];

Table[{i=i+1,
U1[i]=a+(b-a)*RandomReal[],
U2[i]=g+(h-g)*RandomReal[],
U[i]=0+(c-0)*RandomReal[]},{i,0,N1-1,1}];
TableForm[%,TableSpacing->{2,2},
TableHeadings->{None,{"i","U1[i]","U2[i]","U[i]"}}]
```

part B

```
UA1={};
i=0;
While[i<=N1,If[U[i]<norc*(1/(U1[i]*U2[i])),
AppendTo[UA1,U1[i]]];i=i+1];
n1=Length[UA1]

UA2={};
i=0;
While[i<=N2,If[U[i]<norc*(1/(U1[i]*U2[i])),
AppendTo[UA2,U2[i]]];i=i+1];
n2=Length[UA2]

Table[{i=i+1,UA1[[i]]},{i,0,n1-1,1}];
TableForm[%,TableSpacing->{2,2},TableHeadings->{None,{"i","UA1[[i]]"}}]

Table[{i=i+1,UA2[[i]]},{i,0,n2-1,1}];
TableForm[%,TableSpacing->{2,2},TableHeadings->{None,{"i","UA2[[i]]"}}]
```

part C

```
Imc=(1/(n1*n2))*Sum[(((UA1[[i]]+UA2[[j]])^(-y))/
(norc*(1/(UA1[[i]]*UA2[[j]])))),{i,1,n1},{j,1,n2}]
Ini=NIntegrate[(x1+x2)^(-y),{x1,a,b},{x2,g,h}]
error1=(Imc-Ini)*100/Ini
```

We have chosen a probability density function
$$p(x_1, x_2) = norc \ (1/x_1) \ (1/x_2)$$
-------(2.9)

where *norc* is normalization constant. Both the integrand $(x_1 + x_2)^{-y}$ and $p(x_1, x_2) = norc \ (1/x_1) \ (1/x_2)$ are decreasing functions of both x_1 and x_2. Hence maximum value of p is $c = norc \ (1/a)(1/g)$.

In part A of program number 2.3, $U1$ and $U2$ are 2 sets of $N1 = N2$ uniform random numbers for x_1 and x_2 respectively in the interval a to b and g to h respectively. U's are $N_1 = N_2$ uniform random numbers in interval 0 to c.

Part B of program number 2.3 uses the random numbers U_1, U_2 and U according to acceptance-rejection sampling method and produces 2 sets of $n_1 = n_2$ accepted random variates $UA1$ and $UA2$ for x_1 and x_2 respectively shown in Table 2.7 and Table 2.8 respectively. Acceptance-rejection sampling technique uses the condition $U < norc \ (1/U1)(1/U2)$. If the condition is satisfied, the corresponding values of $U1$ and $U2$ are accepted as random variates which we call $UA1[[i]]$ and $UA2[[i]]$ respectively for further calculations.

Table 2.7 Showing accepted values of random variate $U1$ obtained using part B of program number 2.3. We have 4 accepted values out of 200 values of $U1$.

i	UA1[[i]]
1	2.68697
2	2.21476
3	2.49491
4	4.01212

Table 2.8 Showing accepted values of random variate $U2$ obtained using part B of program number 2.3. We have 4 accepted values out of 200 values of $U2$.

i	UA2[[i]]
1	4.79884
2	20.4704
3	3.9611
4	3.79737

Using part C of program number 2.3, we evaluate the integral I_3 using the equation

$$I_3 = \frac{1}{n_1 \ n_2} \sum_{i=1}^{n1} \sum_{j=1}^{n2} \frac{F(x_{1i}, x_{2j})}{p(x_{1i}, x_{2j})} = \frac{1}{n_1 \ n_2} \sum_{i=1}^{n1} \sum_{j=1}^{n2} \frac{(x_{1i} + x_{2j})^{-y}}{norc \ (1/x_{1i})(1/x_{2j})}$$

as $\qquad I_3 = \frac{1}{n_1 \ n_2} \sum_{i=1}^{n1} \sum_{j=1}^{n2} \frac{(\ UA1[[i]] + UA2[[j]] \ )^{-y}}{norc \ (\ 1/UA1[[i]] \ ) \ (\ 1/UA2[[j]] \ )}$ \qquad -------(2.10)

Results are in Table 2.9 as Monte Carlo results I_{mc}.

In part C of the program, we have also calculated the integral as I_{ni} using NIntegrate command. Results are in Table 2.9 as I_{ni}. We have compared I_{ni} with I_{mc} for various values of parameters. The difference has been shown as % error in Table 2.9. Results obtained using program number 2.3 are as shown in Table 2.9.

Table 2.9 Results obtained using the program. Using various values of parameters. We are dealing with

$$I_3 = \int_a^b \int_g^h (x_1 + x_2)^{-y} dx_1 \, dx_2$$

y	a, b	g, h	$N1=$ $N2$	$n1=$ $n2$	$(n1)^2$	Monte Carlo result I_{mc}	result obtained using NIntegrate command I_{ni}	% error
2	1, 11	1, 21	100	3	9	1.35	1.42	-4.60
2	1, 11	1, 21	200	4	16	1.49	1.42	5.17
2	1, 11	1, 21	300	8	64	1.40	1.42	-0.92
2	1, 11	1, 21	500	16	256	1.34	1.42	-5.39
2	1, 11	1, 21	1000	36	1296	1.41	1.42	-0.46
2	1, 21	1, 41	2000	34	1156	2.09	2.01	4.13
2	1, 11	1, 51	2000	43	1849	1.65	1.62	1.99
1.5	1, 11	1, 21	100	3	9	4.04	4.33	-6.76
1.5	1, 11	1, 21	200	4	16	4.38	4.33	0.95
1.5	1, 11	1, 21	500	16	256	4.02	4.33	-7.29
1.5	1, 11	1, 21	800	30	900	4.50	4.33	3.85
1.5	1, 11	1, 21	1000	36	1296	4.28	4.33	-1.33
1.5	1, 21	1, 41	2000	34	1156	7.53	7.53	-0.09
1.5	1, 11	1, 51	2000	43	1849	5.62	5.55	1.29
1	1, 11	1, 21	100	3	9	13.00	14.47	-10.13
1	1, 11	1, 21	300	8	64	12.11	14.47	-16.33
1	1, 11	1, 21	500	16	256	13.25	14.47	-8.41
1	1, 11	1, 21	800	30	900	15.50	14.47	7.12
1	1, 11	1, 21	1000	36	1296	14.20	14.47	-1.84
1	1, 21	1, 41	2000	34	1156	30.54	32.28	-5.39
1	1, 11	1, 51	2000	43	1849	21.80	21.99	-0.85
0.5	1, 11	1, 21	500	16	256	47.54	52.12	-8.79
0.5	1, 11	1, 21	800	30	900	57.22	52.12	9.79
0.5	1, 11	1, 21	1000	36	1296	51.13	52.12	-1.90
0.5	1, 21	1, 41	2000	34	1156	138.14	154.18	-10.41
0.5	1, 11	1, 51	2000	43	1849	95.63	99.29	-3.69

Chapter III

Evaluation of Three-dimensional Definite Integrals

This chapter deals with Monte Carlo evaluation of 4 three-dimensional definite integrals. In each case, a suitable multi-variable probability density function is chosen and acceptance-rejection sampling is used to obtain values of 3 sets of random variates corresponding to the 3 random variables in the probability density function. The integrals are evaluated using the 3 sets of random variates. Programs written in Mathematica have been used in the sampling as well as in evaluating the integrals. Uses of different parts of the programs have been narrated.

3.1 The 3D definite integrals dealt with in this chapter
This chapter deals with the following 4 definite integrals:

$$I_4 = \int_a^b \int_a^b \int_a^b (x_1 + x_2 + x_3)^y \, dx_1 \, dx_2 \, dx_3 \quad\quad\quad \text{-------(3.1)}$$

$$I_5 = \int_a^b \int_a^b \int_a^b (x_1 + x_2 + x_3)^{-y} \, dx_1 \, dx_2 \, dx_3 \quad\quad\quad \text{-------(3.2)}$$

$$I_6 = \int_a^b \int_a^b \int_a^b \left(\frac{x_1}{x_2 + x_3}\right)^y \, dx_1 \, dx_2 \, dx_3 \quad\quad\quad \text{-------(3.3)}$$

$$I_7 = \int_a^b \int_a^b \int_a^b \left(\frac{x_1}{x_2 + x_3}\right)^{-y} \, dx_1 \, dx_2 \, dx_3 \quad\quad\quad \text{-------(3.4)}$$

where y is positive constant such as 0.5, 2 etc.; limits a and b are also constants such as 1, 5, 11, 21, 31 etc.

3.2 Evaluation of the integral I_4

We now deal with the definite integral $I_4 = \int_a^b \int_a^b \int_a^b (x_1 + x_2 + x_3)^y \, dx_1 \, dx_2 \, dx_3$ where y is positive constant. We first choose a probability density function

$$p(x_1, x_2, x_3) = norc \ (x_1) \ (x_2) \ (x_3) \quad\quad\quad \text{-------(3.5)}$$

where $norc$ is normalization constant. Both $F(x_1, x_2, x_3)$ = $(x_1 + x_2 + x_3)^y$ and $p(x_1, x_2, x_3) = norc \ (x_1) \ (x_2) \ (x_3)$ are increasing functions of x_1, x_2 and x_3. The probability density function can be normalized without running into the problem of evaluating another multi-dimensional definite integral. Maximum value of the probability density function is $c = norc \ (b)(b)(b)$.

Program number 3.1
part A

```
y=2
N1=100;
a=1;
b=21;
norc=1/(NIntegrate[(x1)*(x2)*(x3),{x1,a,b},{x2,a,b},{x3,a,b}])
c=norc*(b)*(b)*(b)
n=654321;
SeedRandom[n];
Table[{i=i+1,
```

```
U1[i]=a+(b-a)*RandomReal[],
U2[i]=a+(b-a)*RandomReal[],
U3[i]=a+(b-a)*RandomReal[],
u[i]=RandomReal[];U[i]=0+(c-0)*u[i]},{i,0,N1-1,1}];
TableForm[%,TableSpacing->{2,2},
TableHeadings->{None,{"i","U1[i]","U2[i]","U3[i]","U[i]"}}]
```

part B
```
UA1={};
i=0;
While[i<=N1,If[U[i]<norc*(U1[i]*U2[i]*U3[i]),
AppendTo[UA1,U1[i]]];i=i+1];
n1=Length[UA1]
```

```
UA2={};
i=0;
While[i<=N1,If[U[i]<norc*(U1[i]*U2[i]*U3[i]),
AppendTo[UA2,U2[i]]];i=i+1];
n2=Length[UA2]
```

```
UA3={};
i=0;
While[i<=N1,If[U[i]<norc*(U1[i]*U2[i]*U3[i]),
AppendTo[UA3,U3[i]]];i=i+1];
n3=Length[UA3]
```

```
Table[{i=i+1,UA1[[i]],UA2[[i]],UA3[[i]]},{i,0,n1-1,1}];
TableForm[%,TableSpacing->{2,2},
TableHeadings->{None,{"i","UA1[[i]]","UA2[[i]]","UA3[[i]]"}}]
```

part C
```
Imc=(1/(n1*n2*n3))*Sum[(((UA1[[i]]+UA2[[j]]+UA3[[k]])^y)/
(norc*(UA1[[i]]*UA2[[j]]*UA3[[k]])),{i,1,n1},{j,1,n2},{k,1,n3}]
Ini=NIntegrate[(x1+x2+x3)^y,{x1,a,b},{x2,a,b},{x3,a,b}]
error1=(Imc-Ini)*100/Ini
```

To evaluate the 3D definite integral I_4, we have written program number 3.1. In part A of program number 3.1, $U1$, $U2$ and $U3$ are 3 sets of $N1$ uniform random numbers for x_1, x_2 and x_3 respectively in the interval a to b. u and U are 2 sets of $N1$ uniform random numbers in interval 0 to 1 and 0 to c respectively. Table 3.1 shows these random numbers $U1$, $U2$, $U3$, U for $N1 = 100$.

Table 3.1 Showing values of random numbers obtained using part A of program number 3.1 for $N1 = 100$. The program performs acceptance-rejection sampling using these random numbers.

i	$U1[i]$	$U2[i]$	$U3[i]$	$U[i]$
1	18.2366	9.5744	18.1920	0.0000
2	16.1932	10.2171	9.3782	0.0004
3	8.5982	4.3739	4.7988	0.0000
4	7.6694	9.5138	8.7210	0.0002
5	12.7629	5.6443	14.4863	0.0008
6	18.5177	8.3163	6.8907	0.0006
7	19.2079	19.5518	2.0081	0.0003
8	4.0208	4.5228	8.5926	0.0005
9	8.1724	9.4040	12.2827	0.0001
10	2.5478	18.8487	20.6957	0.0004
...	...	...	...	...
91	10.9074	8.1776	13.5069	0.0000
92	19.0230	9.4458	1.9304	0.0005
93	5.3868	10.7936	4.4265	0.0002
94	3.0048	14.6937	10.3036	0.0002
95	19.6323	6.5132	16.6968	0.0008
96	8.9498	8.5839	5.6998	0.0007
97	4.3864	6.5411	7.6125	0.0003
98	18.3030	15.1030	16.9062	0.0008
99	10.7510	7.0896	12.1794	0.0006
100	3.9776	1.9979	12.2301	0.0005

Part B of program number 3.1 uses the random numbers of Table 3.1 according to acceptance-rejection sampling method and produces 3 sets of $n_1 = n_2 = n_3 = 16$ accepted random variates UA1, UA2 and UA3 for x_1, x_2 and x_3 respectively shown in Table 3.2 for $N1 = 100$. Acceptance-rejection sampling technique uses the condition $U < norc$ $(U1)(U2)(U3)$. If the condition is satisfied, the $U1$, $U2$ and $U3$ are accepted as random variates which we call $UA1[[i]]$, $UA2[[i]]$ and $UA3[[i]]$ respectively for further calculations.

Table 3.2 Showing accepted values of random variates obtained using part B of program number 3.1. We have 16 accepted values out of 100 for $U1$, $U2$ and $U3$.

i	UA1[[i]]	UA2[[i]]	UA3[[i]]
1	18.2366	9.5744	18.1920
2	9.2673	5.9028	12.1303
3	14.5404	15.5312	9.7957
4	19.5874	2.9414	18.3567
5	19.4975	13.6215	13.5519
6	3.9086	19.8327	8.8593
7	15.6992	10.7704	10.5620
8	18.4490	6.7574	11.2425
9	6.6558	19.8923	12.7691

10	15.5983	17.2234	19.5775
11	18.3454	11.4777	14.4601
12	10.5345	13.8822	13.6226
13	11.3372	8.1394	19.1559
14	15.2577	8.1911	18.8642
15	11.1724	16.8408	14.2183
16	10.9074	8.1776	13.5069

Using part C of program number 3.1, we evaluate the integral I_4 using the following equation

$$I_4 = \frac{1}{n_1\, n_2\, n_3} \sum_{i=1}^{n1} \sum_{j=1}^{n2} \sum_{k=1}^{n3} \frac{F(x_{1i}, x_{2j}, x_{3k})}{p(x_{1i}, x_{2j}, x_{3k})} = \frac{1}{n_1\, n_2\, n_3} \sum_{i=1}^{n1} \sum_{j=1}^{n2} \sum_{k=1}^{n3} \frac{(x_{1i} + x_{2j} + x_{3k})^y}{norc\ (x_{1i})(x_{2j})(x_{3k})}$$

as

$$I_4 = \frac{1}{n_1\, n_2\, n_3} \sum_{i=1}^{n1} \sum_{j=1}^{n2} \sum_{k=1}^{n3} \frac{(\ UA1[[i]] + UA2[[j]]\ + UA3[[k]])^y}{norc\ (\ UA1[[i]]\ )\ (\ UA2[[j]]\ )\ (\ UA3[[k]]\ )} \qquad \text{--------(3.6)}$$

Results are in Table 3.3 as Monte Carlo result I_{mc}.

In part C of program number 3.1, we have also calculated the integral as I_{ni} using NIntegrate command. Results are in Table 3.3 as I_{ni}. We have compared I_{ni} with I_{mc} for various values of parameters and of $N1$. The difference is denoted as error1 which is % error shown in Table 3.3.

Table 3.3 Results of part C of program number 3.1 for various values of parameters. We are dealing with

$$I_4 = \int_a^b \int_a^b \int_a^b (x_1 + x_2 + x_3)^y\, dx_1\ dx_2\ dx_3$$

i	y	a	b	N1	n1	n1^3	Monte Carlo result I_{mc}	result obtained using NIntegrate command I_{ni}	% error
1	2	1	5	50	11	1331	5905.91	5440.00	8.56
2	2	1	11	50	7	343	370079.00	349000.00	6.04
3	2	1	21	100	16	4096	9.21*10^6	9.51*10^6	-3.21
4	2	1	31	100	15	3375	6.35*10^7	6.83*10^7	-6.94
5	2	1	41	100	15	3375	2.61*10^8	2.80*10^8	-6.49
6	2	1	51	100	15	3375	7.87*10^8	8.39*10^8	-6.18
7	2	1	71	100	14	2744	4.22*10^9	4.42*10^9	-4.61
8	2	1	101	100	14	2744	2.48*10^10	2.59*10^10	-4.24
9	0.5	1	11	100	17	4913	4592.49	4198.64	9.38
10	0.5	1	11	120	20	8000	4164.93	4198.64	-0.80
11	0.5	1	21	150	24	13824	35286.20	45379.50	-22.24
12	0.5	1	21	165	29	24389	38352.70	45379.50	-15.48
13	0.5	1	21	180	31	29791	43102.30	45379.50	-5.02
14	0.5	1	31	200	34	39304	174794.00	184545.00	-5.28
15	0.5	1	41	200	34	39304	485039.00	500905.00	-3.17
16	0.5	1	51	200	34	39304	1.07*10^6	1.09*10^6	-1.70
17	0.5	1	71	200	33	35937	3.56*10^6	3.51*10^6	1.23
18	0.5	1	101	200	33	35937	1.25*10^7	1.22*10^7	2.91

Table 3.3 displays a survey of Monte Carlo evaluation of the 3D definite integration I_4 for different values of various parameters. We note that, while using Monte Carlo method, % error is generally below 10%. If the error is more than 10%, the error can usually be lowered by raising number of terms $n1^3$ in the summation in equation (3.6) by raising value of $N1$.

3.3 Evaluation of the integral I_5

We now deal with the definite integral $I_5 = \int_a^b \int_a^b \int_a^b (x_1 + x_2 + x_3)^{-y} dx_1 \ dx_2 \ dx_3$ where y is positive constant. We first choose a probability density function

$$p(x_1, x_2, x_3) = norc \ (1/x_1) \ (1/x_2) \ (1/x_3) \qquad \text{-------(3.7)}$$

where   *norc*   is   normalization   constant.   Both $F(x_1, x_2, x_3)$ = $(x_1 + x_2 + x_3)^{-y}$ and $p(x_1, x_2, x_3) = norc \ (1/x_1) \ (1/x_2) \ (1/x_3)$ are decreasing functions of x_1, x_2 and x_3. The probability density function can be normalized without running into the problem of evaluating another multi-dimensional definite integral. Maximum value of the probability density function is $c = norc \ (1/a)(1/a)(1/a)$.

Program number 3.2
part A

```
y=2

N1=150;

a=1;

b=5;

norc=1/(NIntegrate[(1/x1)*(1/x2)*(1/x3),{x1,a,b},{x2,a,b},{x3,a,b}])

c=norc*(1/a)*(1/a)*(1/a)

n=654321;

SeedRandom[n];

Table[{i=i+1,

U1[i]=a+(b-a)*RandomReal[],

U2[i]=a+(b-a)*RandomReal[],

U3[i]=a+(b-a)*RandomReal[],

u[i]=RandomReal[];U[i]=0+(c-0)*u[i]},{i,0,N1-1,1}];

TableForm[%,TableSpacing->{2,2},

TableHeadings->{None,{"i","U1[i]","U2[i]","U3[i]","U[i]"}}]
```

part B

```
UA1={};

i=0;

While[i<=N1,If[U[i]<norc*(1/(U1[i]*U2[i]*U3[i])),

AppendTo[UA1,U1[i]]];i=i+1];

n1=Length[UA1]
```

```
UA2={};
i=0;
While[i<=N1,If[U[i]<norc*(1/(U1[i]*U2[i]*U3[i])),
AppendTo[UA2,U2[i]]];i=i+1];
n2=Length[UA2]


UA3={};
i=0;
While[i<=N1,If[U[i]<norc*(1/(U1[i]*U2[i]*U3[i])),
AppendTo[UA3,U3[i]]];i=i+1];
n3=Length[UA3]


Table[{i=i+1,UA1[[i]],UA2[[i]],UA3[[i]]},{i,0,n1-1,1}];
TableForm[%,TableSpacing->{2,2},
TableHeadings->{None,{"i","UA1[[i]]","UA2[[i]]","UA3[[i]]"}}]
```

part C

```
Imc=(1/(n1*n2*n3))*Sum[((( UA1[[i]]+UA2[[j]]+UA3[[k]] )^(-y))/
(norc*(1/( UA1[[i]]*UA2[[j]]*UA3[[k]]))),{i,1,n1},{j,1,n2},{k,1,n3}]
Ini=NIntegrate[(x1+x2+x3)^(-y),{x1,a,b},{x2,a,b},{x3,a,b}]
error1=(Imc-Ini)*100/Ini
```

To evaluate the 3D definite integral I_5, we have written program number 3.2. In part A of program number 3.2, $U1$, $U2$ and $U3$ are 3 sets of $N1$ uniform random numbers for x_1, x_2 and x_3 respectively in the interval a to b. u and U are 2 sets of $N1$ uniform random numbers in interval 0 to 1 and 0 to c respectively. Table 3.4 shows these random numbers $U1$, $U2$, $U3$, U for $N1 = 150$.

Table 3.4 Showing values of random numbers obtained using part A of program number 3.2 for $N1 = 150$. The program performs acceptance-rejection sampling using these random numbers.

i	$U1[i]$	$U2[i]$	$U3[i]$	$U[i]$
1	4.4473	2.7149	4.4384	0.0082
2	4.0386	2.8434	2.6756	0.1138
3	2.5196	1.6748	1.7598	0.0087
4	2.3339	2.7028	2.5442	0.0586
5	3.3526	1.9289	3.6973	0.2180
6	4.5035	2.4633	2.1781	0.1612
7	4.6416	4.7104	1.2016	0.0842
8	1.6042	1.7046	2.5185	0.1257
9	2.4345	2.6808	3.2565	0.0329
10	1.3096	4.5698	4.9391	0.1164
...	...	...	...	...
141	1.1345	1.6135	2.1328	0.1667
142	2.6478	2.1495	3.5014	0.1990

143	3.4921	3.7595	1.0923	0.0116
144	4.0813	4.5666	2.9630	0.0768
145	4.5780	4.3445	3.4588	0.1691
146	4.5956	3.4454	2.5903	0.2374
147	3.4967	1.5258	1.4967	0.0686
148	2.2049	1.5595	1.0150	0.0464
149	1.1941	4.5172	2.0182	0.2307
150	4.5239	4.3763	4.0730	0.2109

Part B of program number 3.2 uses the random numbers of Table 3.4 according to acceptance-rejection sampling method and produces 3 sets of $n_1 = n_2 = n_3 = 10$ accepted random variates UA1, UA2 and UA3 for x_1, x_2 and x_3 respectively shown in Table 3.5 for $N1 = 150$. Acceptance-rejection sampling technique uses the condition $U < norc$ $(1/U1)(1/U2)(1/U3)$. If the condition is satisfied, the $U1$, $U2$ and $U3$ are accepted as random variates which we call $UA1[[i]]$, $UA2[[i]]$ and $UA3[[i]]$ respectively for further calculations.

Table 3.5 Showing accepted values of random variates obtained using part B of program number 3.2. We have 10 accepted values out of 150 for $U1$, $U2$ and $U3$.

i	UA1[[i]]	UA2[[i]]	UA3[[i]]
1	2.5196	1.6748	1.7598
2	1.4317	2.9516	1.6620
3	2.6535	1.9806	3.2261
4	4.4898	2.1515	3.0485
5	1.1996	4.4320	1.8900
6	2.9069	3.5764	3.5245
7	2.9815	2.4355	3.5014
8	2.1898	1.0296	3.1308
9	3.4921	3.7595	1.0923
10	2.2049	1.5595	1.0150

Using part C of program number 3.2, we evaluate the integral I_5 using the following equation

$$I_5 = \frac{1}{n_1\,n_2\,n_3} \sum_{i=1}^{n1} \sum_{j=1}^{n2} \sum_{k=1}^{n3} \frac{F(x_{1i}, x_{2j}, x_{3k})}{p(x_{1i}, x_{2j}, x_{3k})}$$

$$= \frac{1}{n_1\,n_2\,n_3} \sum_{i=1}^{n1} \sum_{j=1}^{n2} \sum_{k=1}^{n3} \frac{(x_{1i} + x_{2j} + x_{3k})^{-y}}{norc\,(1/x_{1i})(1/x_{2j})(1/x_{3k})}$$

as

$$I_5 = \frac{1}{n_1\,n_2\,n_3} \sum_{i=1}^{n1} \sum_{j=1}^{n2} \sum_{k=1}^{n3} \frac{(\,UA1[[i]] + UA2[[j]] + UA3[[k]])^{-y}}{norc\,(\,1/UA1[[i]]\,)\,(\,1/UA2[[j]]\,)\,(\,1/UA3[[k]]\,)} \qquad \text{-------(3.8)}$$

Results are in Table 3.6 as Monte Carlo result I_{mc}.

In part C of program number 3.2, we have also calculated the integral as I_{ni} using NIntegrate command. Results are in Table 3.6 as I_{ni}. We have compared I_{ni} with I_{mc} for various values of parameters and of $N1$. The difference is denoted as error1 which is % error shown in Table 3.6.

Table 3.6 Results of part C of program number 3.2 for various values of parameters. We are dealing with

$$I_5 = \int_a^b \int_a^b \int_a^b (x_1 + x_2 + x_3)^{-y} dx_1 \ dx_2 \ dx_3$$

i	y	a	b	$N1$	$n1$	$n1\text{\textasciicircum}3$	Monte Carlo result I_{mc}	result obtained using NIntegrate command I_{ni}	% error
1	2	1	5	150	10	1000	1.00	0.94	5.82
2	2	1	11	1000	10	1000	4.33	4.23	2.51
3	2	1	21	3500	13	2197	10.50	11.12	-5.57
4	2	1	31	7000	10	1000	19.11	18.66	2.40
5	0.5	1	5	200	15	3375	20.51	21.77	-5.79
6	0.5	1	21	4000	16	4096	1418.14	1452.93	-2.39
7	0.5	1	31	8000	11	1331	4209.34	4080.51	3.16

Table 3.6 displays a survey of Monte Carlo evaluation of the 3D definite integration I_5 for different values of various parameters. We note that, while using Monte Carlo method, % error is generally below 10%. If the error is more than 10%, the error can usually be lowered by raising number of terms $n1^3$ in the summation in equation (3.8) by raising value of $N1$.

3.4 Evaluation of the integral I_6

We now deal with the definite integral $I_6 = \int_a^b \int_a^b \int_a^b \left(\dfrac{x_1}{x_2 + x_3} \right)^y dx_1 \ dx_2 \ dx_3$ where y is positive constant. We first choose a probability density function

$$p(x_1, x_2, x_3) = norc \ (x_1) \ (1/x_2) \ (1/x_3) \qquad \text{-------(3.9)}$$

where $norc$ is normalization constant. Both $F(x_1, x_2, x_3) = \left(\dfrac{x_1}{x_2 + x_3} \right)^y$ and

$p(x_1, x_2, x_3) = norc \ (x_1) \ (1/x_2) \ (1/x_3)$ are increasing functions of x_1 and decreasing functions of x_2 and x_3. The probability density function can be normalized without running into the problem of evaluating another multi-dimensional definite integral. Maximum value of the probability density function is $c = norc \ (b)(1/a)(1/a)$.

Program number 3.3
part A

y=0.5

N1=150;

a=1;

b=5;

norc=1/(NIntegrate[(x1)*(1/x2)*(1/x3),{x1,a,b},{x2,a,b},{x3,a,b}])

c=norc*(b)*(1/a)*(1/a)

n=654321;

SeedRandom[n];

```
Table[{i=i+1,
U1[i]=a+(b-a)*RandomReal[],
U2[i]=a+(b-a)*RandomReal[],
U3[i]=a+(b-a)*RandomReal[],
u[i]=RandomReal[];U[i]=0+(c-0)*u[i]},{i,0,N1-1,1}];
TableForm[%,TableSpacing->{2,2},
TableHeadings->{None,{"i","U1[i]","U2[i]","U3[i]", "U[i]"}}]
```

part B
```
UA1={};
i=0;
While[i<=N1,If[U[i]<norc*(U1[i])*(1/U2[i])*(1/U3[i]),
AppendTo[UA1,U1[i]]];i=i+1];
n1=Length[UA1]
```

```
UA2={};
i=0;
While[i<=N1,If[U[i]<norc*(U1[i])*(1/U2[i])*(1/U3[i]),
AppendTo[UA2,U2[i]]];i=i+1];
n2=Length[UA2]
```

```
UA3={};
i=0;
While[i<=N1,If[U[i]<norc*(U1[i])*(1/U2[i])*(1/U3[i]),
AppendTo[UA3,U3[i]]];i=i+1];
n3=Length[UA3]
```

```
Table[{i=i+1,UA1[[i]],UA2[[i]],UA3[[i]]},{i,0,n1-1,1}];
TableForm[%,TableSpacing->{2,2},
TableHeadings->{None,{"i","UA1[[i]]","UA2[[i]]","UA3[[i]]"}}]
```

part C
```
Imc=(1/(n1*n2*n3))*Sum[(( UA1[[i]]/(UA2[[j]]+UA3[[k]]))^y)/
(norc*(UA1[[i]])*(1/UA2[[j]])*(1/UA3[[k]])),{i,1,n1},{j,1,n2},{k,1,n3}]
Ini=NIntegrate[(x1/(x2+x3))^y,{x1,a,b},{x2,a,b},{x3,a,b}]
error1=(Imc-Ini)*100/Ini
```

To evaluate the 3D definite integral I_6, we have written program number 3.3. In part A of program number 3.3, $U1$, $U2$ and $U3$ are 3 sets of $N1$ uniform random numbers for x_1, x_2 and x_3 respectively in the interval a to b. u and U are 2 sets of $N1$ uniform random numbers in interval 0 to 1 and 0 to c respectively. Table 3.7 shows these random numbers $U1$, $U2$, $U3$, U for $N1$ = 150.

Table 3.7 Showing values of random numbers obtained using part A of program number 3.3 for $N1$ = 150. The program performs acceptance-rejection sampling using these random numbers.

i	U1[i]	U2[i]	U3[i]	U[i]
1	4.4473	2.7149	4.4384	0.0055
2	4.0386	2.8434	2.6756	0.0763
3	2.5196	1.6748	1.7598	0.0059
4	2.3339	2.7028	2.5442	0.0393
5	3.3526	1.9289	3.6973	0.1462
6	4.5035	2.4633	2.1781	0.1081
7	4.6416	4.7104	1.2016	0.0565
8	1.6042	1.7046	2.5185	0.0843
9	2.4345	2.6808	3.2565	0.0221
10	1.3096	4.5698	4.9391	0.0781
...	...	...	...	...
141	1.1345	1.6135	2.1328	0.1118
142	2.6478	2.1495	3.5014	0.1335
143	3.4921	3.7595	1.0923	0.0078
144	4.0813	4.5666	2.9630	0.0515
145	4.5780	4.3445	3.4588	0.1134
146	4.5956	3.4454	2.5903	0.1592
147	3.4967	1.5258	1.4967	0.0460
148	2.2049	1.5595	1.0150	0.0311
149	1.1941	4.5172	2.0182	0.1547
150	4.5239	4.3763	4.0730	0.1414

Part B of program number 3.3 uses the random numbers of Table 3.7 according to acceptance-rejection sampling method and produces 3 sets of $n_1 = n_2 = n_3 = 18$ accepted random variates UA1, UA2 and UA3 for x_1, x_2 and x_3 respectively shown in Table 3.8 for $N1$ = 150. Acceptance-rejection sampling technique uses the condition $U < norc$ $(U1)(1/U2)(1/U3)$. If the condition is satisfied, the $U1$, $U2$ and $U3$ are accepted as random variates which we call $UA1[[i]]$, $UA2[[i]]$ and $UA3[[i]]$ respectively for further calculations.

Table 3.8 Showing accepted values of random variates obtained using part B of program number 3.3. We have 18 accepted values out of 150 for $U1$, $U2$ and $U3$.

i	UA1[[i]]	UA2[[i]]	UA3[[i]]
1	4.4473	2.7149	4.4384
2	2.5196	1.6748	1.7598
3	4.6097	1.0200	1.8731
4	2.6535	1.9806	3.2261
5	4.7175	1.3883	4.4713
6	3.9398	2.9541	2.9124
7	3.4783	1.5278	2.9043
8	4.4898	2.1515	3.0485
9	1.1996	4.4320	1.8900
10	2.9069	3.5764	3.5245

11	3.8515	2.4382	4.5728
12	2.5589	4.8874	1.1174
13	2.9815	2.4355	3.5014
14	3.8409	2.6352	2.8564
15	2.1898	1.0296	3.1308
16	3.4921	3.7595	1.0923
17	3.4967	1.5258	1.4967
18	2.2049	1.5595	1.0150

Using part C of program number 3.3, we evaluate the integral I_6 using the following equation

$$I_6 = \frac{1}{n_1\,n_2\,n_3} \sum_{i=1}^{n_1} \sum_{j=1}^{n_2} \sum_{k=1}^{n_3} \frac{F(x_{1i}, x_{2j}, x_{3k})}{p(x_{1i}, x_{2j}, x_{3k})}$$

$$= \frac{1}{n_1\,n_2\,n_3} \sum_{i=1}^{n_1} \sum_{j=1}^{n_2} \sum_{k=1}^{n_3} \frac{(x_{1i}/(x_{2j} + x_{3k}))^y}{norc\ (x_{1i})(1/x_{2j})(1/x_{3k})}$$

as

$$I_6 = \frac{1}{n_1\,n_2\,n_3} \sum_{i=1}^{n_1} \sum_{j=1}^{n_2} \sum_{k=1}^{n_3} \frac{(\ UA1[[i]]\ /\ (UA2[[j]]\ + UA3[[k]])\ )^y}{norc\ (\ UA1[[i]]\ )\ (\ 1/UA2[[j]]\ )\ (\ 1/UA3[[k]]\ )} \qquad \text{------(3.10)}$$

Results are in Table 3.9 as Monte Carlo result I_{mc}.

In part C of program number 3.3, we have also calculated the integral as I_{ni} using NIntegrate command. Results are in Table 3.9 as I_{ni}. We have compared I_{ni} with I_{mc} for various values of parameters and of $N1$. The difference is denoted as error1 which is % error shown in Table 3.9.

Table 3.9 Results of part C of program number 3.3 for various values of parameters. We are dealing with

$$I_6 = \int_a^b \int_a^b \int_a^b \left(\frac{x_1}{x_2 + x_3}\right)^y dx_1\,dx_2\,dx_3$$

i	y	a	b	$N1$	$n1$	$n1^3$	Monte Carlo result I_{mc}	result obtained using NIntegrate command I_{ni}	% error
1	2	1	5	80	11	1331	25.12	24.30	3.40
2	2	1	11	300	12	1728	478.04	525.63	-9.05
3	2	1	11	500	16	4096	506.80	525.63	-3.58
4	2	1	21	1500	9	729	4754.56	5405.58	-12.04
5	2	1	21	2000	19	6859	5105.78	5405.58	-5.55
6	0.5	1	5	100	13	2197	54.79	45.76	19.72
7	0.5	1	5	150	18	5832	48.71	45.77	6.43
8	0.5	1	5	300	37	50653	45.43	45.76	-0.73
9	0.5	1	11	1000	26	17576	635.62	721.62	-11.91
10	0.5	1	11	2000	65	274625	656.92	721.62	-8.97
11	0.5	1	21	3000	29	24389	5224.60	5810.65	-10.09
12	0.5	1	21	4000	41	68921	5854.20	5810.65	0.75
13	0.5	1	31	10000	66	287496	15296.90	19673.60	-22.25
14	0.5	1	31	15000	111	1367631	17418.60	19673.60	-11.46
15	0.5	1	31	20000	152	3511808	19684.10	19673.60	0.05

Table 3.9 displays a survey of Monte Carlo evaluation of the 3D definite integration I_6 for different values of various parameters. We note that, while using Monte Carlo method, % error is generally below 10%. If the error is more than 10%, the error can usually be lowered by raising number of terms $n1^3$ in the summation in equation (3.10) by raising value of $N1$.

3.5 Evaluation of the integral I_7

We now deal with the definite integral $I_7 = \int\limits_a^b \int\limits_a^b \int\limits_a^b \left(\dfrac{x_1}{x_2 + x_3} \right)^{-y} dx_1 \, dx_2 \, dx_3$ where y is positive constant. We first choose a probability density function

$$p(x_1, x_2, x_3) = norc \ (1/x_1) \ (x_2) \ (x_3) \qquad\qquad \text{-------(3.11)}$$

where $norc$ is normalization constant. Both $F(x_1, x_2, x_3) = \left(\dfrac{x_1}{x_2 + x_3} \right)^{-y}$ and

$p(x_1, x_2, x_3) = norc \ (1/x_1) \ (x_2) \ (x_3)$ are decreasing functions of x_1 and increasing functions of x_2 and x_3. The probability density function can be normalized without running into the problem of evaluating another multi-dimensional definite integral. Maximum value of the probability density function is $c = norc \ (1/a)(b)(b)$.

Program number 3.4
part A

```
y=0.5
N1=100;
a=1;
b=5;
norc=1/(NIntegrate[(1/x1)*(x2)*(x3),{x1,a,b},{x2,a,b},{x3,a,b}])
c=norc*(1/a)*(b)*(b)
n=654321;
SeedRandom[n];
Table[{i=i+1,
U1[i]=a+(b-a)*RandomReal[],
U2[i]=a+(b-a)*RandomReal[],
U3[i]=a+(b-a)*RandomReal[],
u[i]=RandomReal[];U[i]=0+(c-0)*u[i]},{i,0,N1-1,1}];
TableForm[%,TableSpacing->{2,2},
TableHeadings->{None,{"i","U1[i]","U2[i]","U3[i]","U[i]"}}]
```

part B

```
UA1={};
i=0;
While[i<=N1,If[U[i]<norc*(1/U1[i])*(U2[i])*(U3[i]),
AppendTo[UA1,U1[i]]];i=i+1;
```

n1=Length[UA1]

UA2={};

i=0;

While[i<=N1,If[U[i]<norc*(1/U1[i])*(U2[i])*(U3[i]),

AppendTo[UA2,U2[i]]];i=i+1];

n2=Length[UA2]

UA3={};

i=0;

While[i<=N1,If[U[i]<norc*(1/U1[i])*(U2[i])*(U3[i]),

AppendTo[UA3,U3[i]]];i=i+1];

n3=Length[UA3]

Table[{{i=i+1,UA1[[i]],UA2[[i]],UA3[[i]]},{i,0,n1-1,1}];

TableForm[%,TableSpacing->{2,2},

TableHeadings->{None,{"i","UA1[[i]]","UA2[[i]]","UA3[[i]]"}}]

part C

Imc=(1/(n1*n2*n3))*Sum[(((UA1[[i]]/(UA2[[j]]+UA3[[k]]))^(-y))/

(norc*(1/UA1[[i]])*(UA2[[j]])*(UA3[[k]])),{i,1,n1},{j,1,n2},{k,1,n3}]

Ini=NIntegrate[(x1/(x2+x3))^(-y),{x1,a,b},{x2,a,b},{x3,a,b}]

error1=(Imc-Ini)*100/Ini

To evaluate the 3D definite integral I_7, we have written program number 3.4. In part A of program number 3.4, $U1$, $U2$ and $U3$ are 3 sets of $N1$ uniform random numbers for x_1, x_2 and x_3 respectively in the interval a to b. u and U are 2 sets of $N1$ uniform random numbers in interval 0 to 1 and 0 to c respectively. Table 3.10 shows these random numbers $U1$, $U2$, $U3$, U for $N1 = 100$.

Table 3.10 Showing values of random numbers obtained using part A of program number 3.4 for $N1 = 100$. The program performs acceptance-rejection sampling using these random numbers.

i	U1[i]	U2[i]	U3[i]	U[i]
1	4.4473	2.7149	4.4384	0.0037
2	4.0386	2.8434	2.6756	0.0512
3	2.5196	1.6748	1.7598	0.0039
4	2.3339	2.7028	2.5442	0.0264
5	3.3526	1.9289	3.6973	0.0980
6	4.5035	2.4633	2.1781	0.0725
7	4.6416	4.7104	1.2016	0.0379
8	1.6042	1.7046	2.5185	0.0565

9	2.4345	2.6808	3.2565	0.0148
10	1.3096	4.5698	4.9391	0.0524
...	...	...	...	...
91	2.9815	2.4355	3.5014	0.0028
92	4.6046	2.6892	1.1861	0.0649
93	1.8774	2.9587	1.6853	0.0243
94	1.4010	3.7387	2.8607	0.0193
95	4.7265	2.1026	4.1394	0.1016
96	2.5900	2.5168	1.9400	0.0829
97	1.6773	2.1082	2.3225	0.0331
98	4.4606	3.8206	4.1812	0.0970
99	2.9502	2.2179	3.2359	0.0805
100	1.5955	1.1996	3.2460	0.0662

Part B of program number 3.4 uses the random numbers of Table 3.10 according to acceptance-rejection sampling method and produces 3 sets of $n_1 = n_2 = n_3 = 19$ accepted random variates UA1, UA2 and UA3 for x_1, x_2 and x_3 respectively shown in Table 3.11 for $N1 = 100$. Acceptance-rejection sampling technique uses the condition $U <$ norc $(1/U1)(U2)(U3)$. If the condition is satisfied, the $U1$, $U2$ and $U3$ are accepted as random variates which we call $UA1[[i]]$, $UA2[[i]]$ and $UA3[[i]]$ respectively for further calculations.

Table 3.11 Showing accepted values of random variates obtained using part B of program number 3.4. We have 19 accepted values out of 100 for $U1$, $U2$ and $U3$.

i	UA1[[i]]	UA2[[i]]	UA3[[i]]
1	4.4473	2.7149	4.4384
2	2.5196	1.6748	1.7598
3	2.4345	2.6808	3.2565
4	1.3096	4.5698	4.9391
5	1.4317	2.9516	1.6620
6	2.4176	2.4093	2.5579
7	2.6535	1.9806	3.2261
8	2.2399	3.9457	2.0584
9	1.5817	4.7665	2.5719
10	3.9398	2.9541	2.9124
11	4.4898	2.1515	3.0485
12	2.1312	4.7785	3.3538
13	3.9197	4.2447	4.7155
14	1.1996	4.4320	1.8900
15	1.2104	3.3365	2.2631
16	2.9069	3.5764	3.5245
17	3.8515	2.4382	4.5728
18	2.9815	2.4355	3.5014
19	1.4010	3.7387	2.8607

Using part C of program number 3.4, we evaluate the integral I_7 using the following equation

$$I_7 = \frac{1}{n_1\, n_2\, n_3} \sum_{i=1}^{n1} \sum_{j=1}^{n2} \sum_{k=1}^{n3} \frac{F(x_{1i}, x_{2j}, x_{3k})}{p(x_{1i}, x_{2j}, x_{3k})}$$

$$= \frac{1}{n_1\, n_2\, n_3} \sum_{i=1}^{n1} \sum_{j=1}^{n2} \sum_{k=1}^{n3} \frac{(x_{1i}/(x_{2j} + x_{3k}))^{-y}}{norc\ (1/x_{1i})(x_{2j})(x_{3k})}$$

as

$$I_7 = \frac{1}{n_1\, n_2\, n_3} \sum_{i=1}^{n1} \sum_{j=1}^{n2} \sum_{k=1}^{n3} \frac{(\ UA1[[i]]\ /\ (UA2[[j]] + UA3[[k]])\ )^{-y}}{norc\ (\ 1/UA1[[i]]\ )\ (\ UA2[[j]]\ )\ (\ UA3[[k]]\ )}$$ -------------(3.12)

Results are in Table 3.12 as Monte Carlo result I_{mc}.

In part C of program number 3.4, we have also calculated the integral as I_{ni} using NIntegrate command. Results are in Table 3.12 as I_{ni}. We have compared I_{ni} with I_{mc} for various values of parameters and of $N1$. The difference is denoted as error1 which is % error shown in Table 3.12.

Table 3.12 Results of part C of program number 3.4 for various values of parameters. We are dealing with

$$I_7 = \int_a^b \int_a^b \int_a^b \left(\frac{x_1}{x_2 + x_3} \right)^{-y} dx_1\ dx_2\ dx_3$$

i	y	a	b	$N1$	$n1$	$n1\text{^}3$	Monte Carlo result I_{mc}	result obtained using NIntegrate command I_{ni}	% error
1	0.5	1	5	100	19	6859	104.61	95.93	9.05
2	0.5	1	5	150	26	17576	99.81	95.93	4.04
3	0.5	1	11	150	13	2197	1575.53	1579.21	-0.23
4	0.5	1	21	300	11	1331	12523.60	13178.80	-4.97
5	2	1	5	100	19	6859	455.88	494.93	-7.89
6	2	1	11	300	24	13824	13295.20	14606.10	-8.97
7	2	1	11	400	29	24389	14368.80	14606.10	-1.62
8	2	1	21	400	14	2744	202569.00	209778.00	-3.44

Table 3.12 displays a survey of Monte Carlo evaluation of the 3D definite integration I_7 for different values of various parameters. We note that, while using Monte Carlo method, % error is generally below 10%. If the error is more than 10%, the error can usually be lowered by raising number of terms $n1^3$ in the summation in equation (3.12) by raising value of $N1$.

Chapter IV

Evaluation of Five-dimensional Definite Integrals

This chapter deals with Monte Carlo evaluation of 5 five-dimensional definite integrals. In each case, a suitable multi-variable probability density function is chosen and acceptance-rejection sampling is used to obtain values of 5 sets of random variates corresponding to the 5 random variables in the probability density function. The integrals are evaluated using the 5 sets of random variates. Programs written in Mathematica have been used in the sampling as well as in evaluating the integrals. Uses of different parts of the programs have been narrated.

4.1 The 5D definite integrals dealt with in this chapter

This chapter deals with the following 5 definite integrals:

$$I_8 = \int_a^b \int_a^b \int_a^b \int_a^b \int_a^b (x_1 + x_2 + x_3 + x_4 + x_5)^y \, dx_1 \, dx_2 \, dx_3 \, dx_4 \, dx_5 \qquad\text{-------(4.1)}$$

$$I_9 = \int_a^b \int_a^b \int_a^b \int_a^b \int_a^b (x_1 + x_2 + x_3 + x_4 + x_5)^{-y} \, dx_1 \, dx_2 \, dx_3 \, dx_4 \, dx_5 \qquad\text{-------(4.2)}$$

$$I_{10} = \int_a^b \int_a^b \int_a^b \int_a^b \int_a^b \left(\frac{x_1 + x_2}{x_3 + x_4 + x_5}\right)^y \, dx_1 \, dx_2 \, dx_3 \, dx_4 \, dx_5 \qquad\text{-------(4.3)}$$

$$I_{11} = \int_a^b \int_a^b \int_g^h \int_g^h \int_g^h (x_1 + x_2 + x_3 + x_4 + x_5)^y \, dx_1 \, dx_2 \, dx_3 \, dx_4 \, dx_5 \qquad\text{-------(4.4)}$$

$$I_{12} = \int_a^b \int_a^b \int_p^q \int_p^q \int_p^q \left(\frac{x_1 + x_2}{x_3 + x_4 + x_5}\right)^{-y} \, dx_1 \, dx_2 \, dx_3 \, dx_4 \, dx_5 \qquad\text{-------(4.5)}$$

where y is positive constant such as 0.5, 1, 2 etc.; limits a, b , g, h, p and q are also constants such as 1, 5, 11, 21, 31 etc.

4.2 Evaluation of the integral I_8

We now deal with the definite integral I_8:

$$I_8 = \int_a^b \int_a^b \int_a^b \int_a^b \int_a^b (x_1 + x_2 + x_3 + x_4 + x_5)^y \, dx_1 \, dx_2 \, dx_3 \, dx_4 \, dx_5$$

We first choose a probability density function

$$p(x_1, x_2, x_3, x_4, x_5) = norc \ (x_1) \ (x_2) \ (x_3) \ (x_4) \ (x_5) \qquad\text{-------(4.6)}$$

where $norc$ is normalization constant. Both $F(x_1, x_2, x_3, x_4, x_5) = (x_1 + x_2 + x_3 + x_4 + x_5)^y$ and $p(x_1, x_2, x_3, x_4, x_5) = norc \ (x_1) \ (x_2) \ (x_3) \ (x_4) \ (x_5)$ are increasing functions of x_1, x_2, x_3, x_4 and x_5. The probability density function can be normalized without running into the problem of evaluating another multi-dimensional definite integral. Maximum value of the probability density function is $c = norc \ (b)(b)(b)(b)(b)$.

Program number 4.1
part A

y=3;

N1=150;

a=1;

b=5;

norc=1/(NIntegrate[(x1)*(x2)*(x3)*(x4)*(x5),

{x1,a,b},{x2,a,b},{x3,a,b},{x4,a,b},{x5,a,b}])

c=norc*(b)*(b)*(b)*(b)*(b)

n=654321;

```
SeedRandom[n];
Table[{i=i+1,
U1[i]=a+(b-a)*RandomReal[],
U2[i]=a+(b-a)*RandomReal[],
U3[i]=a+(b-a)*RandomReal[],
U4[i]=a+(b-a)*RandomReal[],
U5[i]=a+(b-a)*RandomReal[],
u[i]=RandomReal[];U[i]=0+(c-0)*u[i]},{i,0,N1-1,1}];
TableForm[%,TableSpacing->{2,2},
TableHeadings->{None,{"i","U1[i]","U2[i]","U3[i]","U4[i]","U5[i]" ,"U[i]"}}]
```

part B
```
UA1={};
i=0;
While[i<=N1,If[U[i]<norc*U1[i]*U2[i]*U3[i]*U4[i]*U5[i],
AppendTo[UA1,U1[i]]];i=i+1];
n1=Length[UA1]
```

```
UA2={};
i=0;
While[i<=N1,If[U[i]<norc*U1[i]*U2[i]*U3[i]*U4[i]*U5[i],
AppendTo[UA2,U2[i]]];i=i+1];
n2=Length[UA2]
```

```
UA3={};
i=0;
While[i<=N1,If[U[i]<norc*U1[i]*U2[i]*U3[i]*U4[i]*U5[i],
AppendTo[UA3,U3[i]]];i=i+1];
n3=Length[UA3]
```

```
UA4={};
i=0;
While[i<=N1,If[U[i]<norc*U1[i]*U2[i]*U3[i]*U4[i]*U5[i],
AppendTo[UA4,U4[i]]];i=i+1];
n4=Length[UA4]
```

```
UA5={};
i=0;
```

49

While[i<=N1,If[U[i]<norc*U1[i]*U2[i]*U3[i]*U4[i]*U5[i],

AppendTo[UA5,U5[i]]];i=i+1];

n5=Length[UA5]

Table[{i=i+1,UA1[[i]],UA2[[i]],UA3[[i]],UA4[[i]],UA5[[i]]},{i,0,n1-1,1}];

TableForm[%,TableSpacing->{2,2},

TableHeadings->{None,{"i","UA1[[i]]","UA2[[i]]",

"UA3[[i]]","UA4[[i]]","UA5[[i]]"}}]

part C

Imc=(1/(n1*n2*n3*n4*n5))*

Sum[(((UA1[[i]]+UA2[[j]]+UA3[[k]]+UA4[[l]]+UA5[[m]])^y)/

(norc*UA1[[i]]*UA2[[j]]*UA3[[k]]*UA4[[l]]*UA5[[m]]),

{i,1,n1},{j,1,n2},{k,1,n3},{l,1,n4},{m,1,n5}]

Ini=NIntegrate[(x1+x2+x3+x4+x5)^y,

{x1,a,b},{x2,a,b},{x3,a,b},{x4,a,b},{x5,a,b}]

error1=(Imc-Ini)*100/Ini

To evaluate the 5D definite integral I_8, we have written program number 4.1. In part A of program number 4.1, $U1$, $U2$, $U3$, $U4$ and $U5$ are 5 sets of $N1$ uniform random numbers for x_1, x_2, x_3, x_4, x_5 respectively in the interval a to b. u and U are 2 sets of $N1$ uniform random numbers in interval 0 to 1 and 0 to c respectively. Table 4.1 shows these random numbers $U1$, $U2$, $U3$, $U4$, $U5$ and U for $N1 = 150$.

Table 4.1 Showing values of random numbers obtained using part A of program number 4.1 for $N1 = 150$. The program performs acceptance-rejection sampling using these random numbers.

i	U1[i]	U2[i]	U3[i]	U4[i]	U5[i]	U[i]
1	4.4473	2.7149	4.4384	1.1370	4.0386	0.0058
2	2.6756	2.8983	2.5196	1.6748	1.7598	0.0005
3	2.3339	2.7028	2.5442	1.9779	3.3526	0.0029
4	3.6973	4.6357	4.5035	2.4633	2.1781	0.0084
5	4.6416	4.7104	1.2016	2.4038	1.6042	0.0022
6	2.5185	3.0958	2.4345	2.6808	3.2565	0.0017
7	1.3096	4.5698	4.9391	2.9414	2.2318	0.0062
8	4.7561	3.3692	1.7684	2.5320	3.7016	0.0037
9	4.8949	2.8002	4.1333	4.1510	3.8047	0.0075
10	3.6464	3.0806	4.6097	1.0200	1.8731	0.0030
...	...	...	...	...	...	...
141	4.9109	3.0947	2.3319	2.1203	1.5154	0.0041
142	1.0362	3.3907	1.8926	4.8248	1.5081	0.0116
143	4.8053	1.5675	1.4136	1.7440	3.1148	0.0050
144	3.8149	1.8528	3.8823	2.6794	1.1325	0.0043

145	4.3512	4.6929	2.7364	3.0386	3.5067	0.0031
146	3.5713	2.2385	3.7861	3.4347	3.6262	0.0091
147	1.6265	3.0118	1.1002	4.9895	1.7632	0.0043
148	3.5760	2.5012	2.7049	3.7598	1.6095	0.0009
149	2.8183	4.7227	1.8739	2.7964	4.9038	0.0085
150	4.8792	3.1200	1.2880	3.2137	4.5686	0.0108

Part B of program number 4.1 uses the random numbers of Table 4.1 according to acceptance-rejection sampling method and produces 5 sets of $n_1 = n_2 = n_3 = n_4 = n_5 = 9$ accepted random variates UA1, UA2, UA3, UA4 and UA5 for x_1, x_2, x_3, x_4 and x_5 respectively shown in Table 4.2 for $N1 = 150$. Acceptance-rejection sampling technique uses the condition $U < norc\ (U1)(U2)(U3)(U4)(U5)$. If the condition is satisfied, the $U1$, $U2$, $U3$, $U4$ and $U5$ are accepted as random variates which we call $UA1[[i]]$, $UA2[[i]]$, $UA3[[i]]$, $UA4[[i]]$ and $UA5[[i]]$ respectively for further calculations.

Table 4.2 Showing accepted values of random variates obtained using part B of program number 4.1. We have 9 accepted values out of 150 for $U1$, $U2$, $U3$, $U4$ and $U5$.

i	UA1[[i]]	UA2[[i]]	UA3[[i]]	UA4[[i]]	UA5[[i]]
1	4.8012	4.4695	2.2399	3.9457	2.0584
2	2.4962	4.8226	3.9197	4.2447	4.7155
3	1.7901	4.0983	3.4424	4.9747	2.1109
4	3.2138	4.7940	1.3674	3.7497	3.3221
5	4.6375	3.4714	1.7736	4.3840	1.5407
6	3.5146	4.7395	4.4158	4.6255	2.7139
7	1.9629	4.7139	1.4850	3.7203	2.0990
8	4.3863	3.1763	4.9420	2.2920	4.8039
9	3.7197	4.3186	3.4600	4.6063	2.0794

Using part C of program number 4.1, we evaluate the integral I_8 using the following equation

$$I_8 = \frac{1}{n_1\ n_2\ n_3\ n_4\ n_5} \sum_{i=1}^{n_1} \sum_{j=1}^{n_2} \sum_{k=1}^{n_3} \sum_{l=1}^{n_4} \sum_{m=1}^{n_5} \frac{F(x_{1i}, x_{2j}, x_{3k}, x_{4l}, x_{5m})}{p(x_{1i}, x_{2j}, x_{3k}, x_{4l}, x_{5m})}$$

$$= \frac{1}{n_1\ n_2\ n_3\ n_4\ n_5} \sum_{i=1}^{n_1} \sum_{j=1}^{n_2} \sum_{k=1}^{n_3} \sum_{l=1}^{n_4} \sum_{m=1}^{n_5} \frac{(x_{1i} + x_{2j} + x_{3k} + x_{4l} + x_{5m})^y}{norc\ (x_{1i})(x_{2j})(x_{3k})(x_{4l})(x_{5m})}$$

as

$$I_8 = \frac{1}{n_1\ n_2\ n_3\ n_4\ n_5} \sum_{i=1}^{n_1} \sum_{j=1}^{n_2} \sum_{k=1}^{n_3} \sum_{l=1}^{n_4} \sum_{m=1}^{n_5}$$

--------(4.7)

$$\frac{(\ UA1[[i]] + UA2[[j]] + UA3[[k]] + UA4[[l]] + UA5[[m]]\ )^y}{norc\ (\ UA1[[i]]\ )\ (\ UA2[[j]]\ )\ (\ UA3[[k]]\ )\ (\ UA4[[l]]\ )\ (\ UA5[[m]]\ )}$$

Results are in Table 4.3 as Monte Carlo result I_{mc}.

In part C of program number 4.1, we have also calculated the integral as I_{ni} using NIntegrate command. Results are in Table 4.3 as I_{ni}. We have compared I_{ni} with I_{mc} for various values of parameters and of $N1$. The difference is denoted as error1 which is % error shown in Table 4.3.

Table 4.3 Results of part C of program number 4.1 for various values of parameters. We are dealing with

$$I_8 = \int\limits_a^b \int\limits_a^b \int\limits_a^b \int\limits_a^b \int\limits_a^b (x_1 + x_2 + x_3 + x_4 + x_5)^y \, dx_1 \, dx_2 \, dx_3 \, dx_4 \, dx_5 .$$

i	y	a	b	$N1$	$n1$	$n1\text{^}5$	Monte Carlo result I_{mc}	result obtained using NIntegrate command I_{ni}	% error
1	3	1	5	150	9	59049	3.63*10^6	3.76*10^6	-3.53
2	3	1	11	250	12	248832	3.34*10^9	3.08*10^9	8.65
3	3	1	21	250	9	59049	6.43*10^11	6.20*10^11	3.58
4	2	1	5	300	25	9765625	260190.00	237227.00	9.68
5	2	1	11	300	18	1889568	1.04*10^8	9.42*10^7	10.28
6	2	1	21	300	14	537824	1.18*10^10	1.02*10^10	15.22
7	2	1	21	350	15	759375	1.10*10^10	1.02*10^10	7.49
8	0.5	0.1	1	300	18	1889568	1.07	0.97	9.99
9	0.5	0.1	2	400	17	1419857	62.13	56.33	10.30
10	0.5	0.1	3	400	15	759375	517.69	566.60	-8.63

Table 4.3 displays a survey of Monte Carlo evaluation of the 5D definite integration I_8 for different values of various parameters. We note that, while using Monte Carlo method, % error is generally below 10%. If the error is more than 10%, the error can usually be lowered by raising number of terms $n1^5$ in the summation in equation (4.7) by raising value of $N1$.

4.3 Evaluation of the integral I_9

We now deal with the definite integral I_9.

$$I_9 = \int\limits_a^b \int\limits_a^b \int\limits_a^b \int\limits_a^b \int\limits_a^b (x_1 + x_2 + x_3 + x_4 + x_5)^{-y} \, dx_1 \, dx_2 \, dx_3 \, dx_4 \, dx_5$$

We first choose a probability density function

$$p(x_1, x_2, x_3, x_4, x_5) = norc \ (1/x_1) \ (1/x_2) \ (1/x_3) \ (1/x_4) \ (1/x_5) \qquad \text{-------(4.8)}$$

where $norc$ is normalization constant. Both $F(x_1, x_2, x_3, x_4, x_5) = (x_1 + x_2 + x_3 + x_4 + x_5)^{-y}$ and $p(x_1, x_2, x_3, x_4, x_5) = norc \ (1/x_1) \ (1/x_2) \ (1/x_3) \ (1/x_4) \ (1/x_5)$ are decreasing functions of x_1, x_2, x_3, x_4 and x_5. The probability density function can be normalized without running into the problem of evaluating another multi-dimensional definite integral. Maximum value of the probability density function is $c = norc \ (1/a)(1/a)(1/a)(1/a)(1/a)$.

Program number 4.2
part A

y=2

N1=300;

a=1;

b=3;

norc=1/(NIntegrate[(1/x1)*(1/x2)*(1/x3)*(1/x4)*(1/x5),

{x1,a,b},{x2,a,b},{x3,a,b},{x4,a,b},{x5,a,b}])

c=norc*(1/a)*(1/a)*(1/a)*(1/a)*(1/a)

n=654321;

SeedRandom[n];

```
Table[{i=i+1,
U1[i]=a+(b-a)*RandomReal[],
U2[i]=a+(b-a)*RandomReal[],
U3[i]=a+(b-a)*RandomReal[],
U4[i]=a+(b-a)*RandomReal[],
U5[i]=a+(b-a)*RandomReal[],
u[i]=RandomReal[];U[i]=0+(c-0)*u[i]},{i,0,N1-1,1}];
TableForm[%,TableSpacing->{2,2},
TableHeadings->{None,{"i","U1[i]","U2[i]","U3[i]","U4[i]","U5[i]","U[i]"}}]
```

part B

```
UA1={};
i=0;
While[i<=N1,If[U[i]<norc*(1/(U1[i]*U2[i]*U3[i]*U4[i]*U5[i])),
AppendTo[UA1,U1[i]]];i=i+1];
n1=Length[UA1]
```

```
UA2={};
i=0;
While[i<=N1,If[U[i]<norc*(1/(U1[i]*U2[i]*U3[i]*U4[i]*U5[i])),
AppendTo[UA2,U2[i]]];i=i+1];
n2=Length[UA2]
```

```
UA3={};
i=0;
While[i<=N1,If[U[i]<norc*(1/(U1[i]*U2[i]*U3[i]*U4[i]*U5[i])),
AppendTo[UA3,U3[i]]];i=i+1];
n3=Length[UA3]
```

```
UA4={};
i=0;
While[i<=N1,If[U[i]<norc*(1/(U1[i]*U2[i]*U3[i]*U4[i]*U5[i])),
AppendTo[UA4,U4[i]]];i=i+1];
n4=Length[UA4]
```

```
UA5={};
i=0;
While[i<=N1,If[U[i]<norc*(1/(U1[i]*U2[i]*U3[i]*U4[i]*U5[i])),
```

AppendTo[UA5,U5[i]]];i=i+1;

n5=Length[UA5]

Table[{i=i+1,UA1[[i]],UA2[[i]],UA3[[i]],UA4[[i]],UA5[[i]]},{i,0,n1-1,1}];

TableForm[%,TableSpacing->{2,2},

TableHeadings->{None,{"i","UA1[[i]]","UA2[[i]]","UA3[[i]]" ,"UA4[[i]]","UA5[[i]]"}}]

part C

Imc=(1/(n1*n2*n3*n4*n5))*

Sum[(((UA1[[i]]+UA2[[j]]+UA3[[k]]+UA4[[l]]+UA5[[m]])^(-y))/

(norc*(1/(UA1[[i]]*UA2[[j]]*UA3[[k]]*UA4[[l]]*UA5[[m]])))),

{i,1,n1},{j,1,n2},{k,1,n3},{l,1,n4},{m,1,n5}]

Ini=NIntegrate[(x1+x2+x3+x4+x5)^(-y),

{x1,a,b},{x2,a,b},{x3,a,b},{x4,a,b},{x5,a,b}]

error1=(Imc-Ini)*100/Ini

To evaluate the 5D definite integral I_9, we have written program number 4.2. In part A of program number 4.2, $U1$, $U2$, $U3$, $U4$ and $U5$ are 5 sets of $N1$ uniform random numbers for x_1, x_2, x_3, x_4, x_5 respectively in the interval a to b. u and U are 2 sets of $N1$ uniform random numbers in interval 0 to 1 and 0 to c respectively. Table 4.4 shows these random numbers $U1$, $U2$, $U3$, $U4$, $U5$ and U for $N1 = 300$.

Table 4.4 Showing values of random numbers obtained using part A of program number 4.2 for $N1 = 300$. The program performs acceptance-rejection sampling using these random numbers.

i	U1[i]	U2[i]	U3[i]	U4[i]	U5[i]	U[i]
1	2.7237	1.8574	2.7192	1.0685	2.5193	0.2880
2	1.8378	1.9491	1.7598	1.3374	1.3799	0.0228
3	1.6669	1.8514	1.7721	1.4890	2.1763	0.1451
4	2.3486	2.8179	2.7518	1.7316	1.5891	0.4198
5	2.8208	2.8552	1.1008	1.7019	1.3021	0.1101
6	1.7593	2.0479	1.7172	1.8404	2.1283	0.0857
7	1.1548	2.7849	2.9696	1.9707	1.6159	0.3090
8	2.8781	2.1846	1.3842	1.7660	2.3508	0.1839
9	2.9474	1.9001	2.5667	2.5755	2.4024	0.3710
10	2.3232	2.0403	2.8049	1.0100	1.4365	0.1475
...	...	...	...	...	...	...
291	2.7624	1.1574	1.9690	2.3435	2.8423	0.3950
292	2.0514	1.0546	1.5450	1.2272	1.5468	0.2841
293	1.2947	2.5528	1.1708	1.3801	2.3081	0.0696
294	1.2057	1.3242	2.6280	2.0003	2.9896	0.0847
295	2.0182	2.8582	2.6557	1.5846	2.8095	0.0846
296	2.7876	2.9424	2.7483	2.9868	2.5815	0.2756
297	2.4200	2.9692	2.2050	1.5635	2.6953	0.3588
298	2.8303	2.7082	1.8161	2.8720	1.2518	0.1659
299	2.5110	1.5399	1.8416	2.1195	1.4316	0.0025
300	2.5302	1.4515	2.6766	1.0163	2.8627	0.6230

Part B of program number 4.2 uses the random numbers of Table 4.4 according to acceptance-rejection sampling method and produces 5 sets of $n_1 = n_2 = n_3 = n_4 = n_5 = 12$ accepted random variates UA1, UA2, UA3, UA4 and UA5 for x_1, x_2, x_3, x_4 and x_5 respectively shown in Table 4.5 for $N1 = 300$. Acceptance-rejection sampling technique uses the condition $U < norc\ (1/U1)(1/U2)(1/U3)(1/U4)(1/U5)$. If the condition is satisfied, the $U1$, $U2$, $U3$, $U4$ and $U5$ are accepted as random variates which we call $UA1[[i]]$, $UA2[[i]]$, $UA3[[i]]$, $UA4[[i]]$ and $UA5[[i]]$ respectively for further calculations.

Table 4.5 Showing accepted values of random variates obtained using part B of program number 4.2. We have 12 accepted values out of 300 for $U1$, $U2$, $U3$, $U4$ and $U5$.

i	UA1[[i]]	UA2[[i]]	UA3[[i]]	UA4[[i]]	UA5[[i]]
1	1.8378	1.9491	1.7598	1.3374	1.3799
2	1.3310	1.2437	1.7088	1.7046	1.7789
3	1.1250	1.0198	1.1055	1.2550	1.6873
4	1.4814	2.8570	1.2425	2.3601	1.5495
5	1.3679	1.7734	1.2371	1.1361	2.8238
6	2.1680	2.3468	1.5811	1.5027	1.1538
7	1.4278	2.6465	1.6104	1.9630	2.1074
8	2.4200	1.4300	1.5158	2.5352	1.0601
9	2.3834	1.8795	1.9210	1.2913	2.0643
10	2.1081	2.9655	2.8327	1.4297	1.6025
11	2.6452	1.1681	2.0569	2.7599	1.6969
12	2.5110	1.5399	1.8416	2.1195	1.4316

Using part C of program number 4.2, we evaluate the integral I_9 using the following equation

$$I_9 = \frac{1}{n_1\, n_2\, n_3\, n_4\, n_5} \sum_{i=1}^{n_1} \sum_{j=1}^{n_2} \sum_{k=1}^{n_3} \sum_{l=1}^{n_4} \sum_{m=1}^{n_5} \frac{F(x_{1i}, x_{2j}, x_{3k}, x_{4l}, x_{5m})}{p(x_{1i}, x_{2j}, x_{3k}, x_{4l}, x_{5m})}$$

$$= \frac{1}{n_1\, n_2\, n_3\, n_4\, n_5} \sum_{i=1}^{n_1} \sum_{j=1}^{n_2} \sum_{k=1}^{n_3} \sum_{l=1}^{n_4} \sum_{m=1}^{n_5} \frac{(x_{1i} + x_{2j} + x_{3k} + x_{4l} + x_{5m})^{-y}}{norc\ (1/x_{1i})(1/x_{2j})(1/x_{3k})(1/x_{4l})(1/x_{5m})}$$

as

$$I_9 = \frac{1}{n_1\, n_2\, n_3\, n_4\, n_5} \sum_{i=1}^{n_1} \sum_{j=1}^{n_2} \sum_{k=1}^{n_3} \sum_{l=1}^{n_4} \sum_{m=1}^{n_5}$$

$$\frac{(\ UA1[[i]] + UA2[[j]] + UA3[[k]] + UA4[[l]] + UA5[[m]])^{-y}}{norc\ (\ 1/UA1[[i]]\ )\ (\ 1/UA2[[j]]\ )\ (\ 1/UA3[[k]]\ )\ (\ 1/UA4[[l]]\ )\ (\ 1/UA5[[m]]\ )}$$

--------(4.9)

Results are in Table 4.6 as Monte Carlo result I_{mc}.

In part C of program number 4.2, we have also calculated the integral as I_{ni} using NIntegrate command. Results are in Table 4.6 as I_{ni}. We have compared I_{ni} with I_{mc} for various values of parameters and of $N1$. The difference is denoted as error1 which is % error shown in Table 4.6.

Table 4.6 Results of part C of program number 4.2 for various values of parameters. We are dealing with

$$I_9 = \int\limits_a^b \int\limits_a^b \int\limits_a^b \int\limits_a^b \int\limits_a^b (x_1 + x_2 + x_3 + x_4 + x_5)^{-y} \, dx_1 \, dx_2 \, dx_3 \, dx_4 \, dx_5 \; .$$

i	y	a	b	N1	n1	n1^5	Monte Carlo result I_{mc}	result obtained using NIntegrate command I_{ni}	% error
1	2	1	3	300	12	248832	0.33	0.34	-2.39
2	2	1	5	700	9	59049	5.33	5.02	6.07
3	1	1	3	300	12	248832	3.10	3.26	-4.67
4	1	1	5	700	9	59049	71.80	70.48	1.88
5	0.5	1	3	300	12	248832	9.59	10.18	-5.87
6	0.5	1	5	700	9	59049	265.94	267.53	-0.60

Table 4.6 displays a survey of Monte Carlo evaluation of the 5D definite integration I_9 for different values of various parameters. We note that, while using Monte Carlo method, % error is generally below 10%. If the error is more than 10%, the error can usually be lowered by raising number of terms $n1^5$ in the summation in equation (4.9) by raising value of $N1$.

4.4 Evaluation of the integral I_{10}

We now deal with the definite integral I_{10}.

$$I_{10} = \int\limits_a^b \int\limits_a^b \int\limits_a^b \int\limits_a^b \int\limits_a^b \left(\frac{x_1 + x_2}{x_3 + x_4 + x_5} \right)^y \, dx_1 \, dx_2 \, dx_3 \, dx_4 \, dx_5$$

We first choose a probability density function

$$p(x_1, x_2, x_3, x_4, x_5) = norc \; (x_1) \; (x_2) \; (1/x_3) \; (1/x_4) \; (1/x_5) \qquad \text{-------(4.10)}$$

where *norc* is normalization constant. Both $F(x_1, x_2, x_3, x_4, x_5) = \left(\dfrac{x_1 + x_2}{x_3 + x_4 + x_5} \right)^y$ and

$p(x_1, x_2, x_3, x_4, x_5) = norc \; (x_1) \; (x_2) \; (1/x_3) \; (1/x_4) \; (1/x_5)$ are increasing functions of x_1 and x_2, and decreasing functions of x_3, x_4 and x_5. The probability density function can be normalized without running into the problem of evaluating another multi-dimensional definite integral. Maximum value of the probability density function is $c = norc \; (b)(b)(1/a)(1/a)(1/a)$.

Program number 4.3
part A

y=2

N1=200;

a=1;

b=5;

norc=1/(NIntegrate[(x1)*(x2)*(1/x3)*(1/x4)*(1/x5),

{x1,a,b},{x2,a,b},{x3,a,b},{x4,a,b},{x5,a,b}])

c=norc*(b)*(b)*(1/a)*(1/a)*(1/a)

n=654321;

```
SeedRandom[n];

Table[{i=i+1,
U1[i]=a+(b-a)*RandomReal[],
U2[i]=a+(b-a)*RandomReal[],
U3[i]=a+(b-a)*RandomReal[],
U4[i]=a+(b-a)*RandomReal[],
U5[i]=a+(b-a)*RandomReal[],
u[i]=RandomReal[];U[i]=0+(c-0)*u[i]},{i,0,N1-1,1}];
TableForm[%,TableSpacing->{2,2},
TableHeadings->{None,{"i","U1[i]","U2[i]","U3[i]","U4[i]" ,"U5[i]","U[i]"}}]
```

part B
```
UA1={};
i=0;
While[i<=N1,If[U[i]<norc*(U1[i])*(U2[i])*(1/U3[i])*(1/U4[i])*(1/U5[i]),
AppendTo[UA1,U1[i]]];i=i+1];
n1=Length[UA1]


UA2={};
i=0;
While[i<=N1,If[U[i]<norc*(U1[i])*(U2[i])*(1/U3[i])*(1/U4[i])*(1/U5[i]),
AppendTo[UA2,U2[i]]];i=i+1];
n2=Length[UA2]


UA3={};
i=0;
While[i<=N1,If[U[i]<norc*(U1[i])*(U2[i])*(1/U3[i])*(1/U4[i])*(1/U5[i]),
AppendTo[UA3,U3[i]]];i=i+1];
n3=Length[UA3]


UA4={};
i=0;
While[i<=N1,If[U[i]<norc*(U1[i])*(U2[i])*(1/U3[i])*(1/U4[i])*(1/U5[i]),
AppendTo[UA4,U4[i]]];i=i+1];
n4=Length[UA4]


UA5={};
```

```
i=0;
While[i<=N1,If[U[i]<norc*(U1[i])*(U2[i])*(1/U3[i])*(1/U4[i])*(1/U5[i]),
AppendTo[UA5,U5[i]]];i=i+1];
n5=Length[UA5]


Table[{i=i+1,UA1[[i]],UA2[[i]],UA3[[i]],UA4[[i]],UA5[[i]]},{i,0,n1-1,1}];
TableForm[%,TableSpacing->{2,2},
TableHeadings->{None,{"i","UA1[[i]]","UA2[[i]]","UA3[[i]]" ,"UA4[[i]]","UA5[[i]]"}}]
```

part C

```
Imc=(1/(n1*n2*n3*n4*n5))*
Sum[((((UA1[[i]]+UA2[[j]])/(UA3[[k]]+UA4[[l]]+UA5[[m]]))^y)/
(norc*(UA1[[i]])*(UA2[[j]])*(1/UA3[[k]])*(1/UA4[[l]])*(1/UA5[[m]])),
{i,1,n1},{j,1,n2},{k,1,n3},{l,1,n4},{m,1,n5}]


Ini=NIntegrate[((x1+x2)/(x3+x4+x5))^y,
{x1,a,b},{x2,a,b},{x3,a,b},{x4,a,b},{x5,a,b}]
error1=(Imc-Ini)*100/Ini
```

To evaluate the 5D definite integral I_{10}, we have written program number 4.3. In part A of program number 4.3, $U1$, $U2$, $U3$, $U4$ and $U5$ are 5 sets of $N1$ uniform random numbers for x_1, x_2, x_3, x_4, x_5 respectively in the interval a to b. u and U are 2 sets of $N1$ uniform random numbers in interval 0 to 1 and 0 to c respectively. Table 4.7 shows these random numbers $U1$, $U2$, $U3$, $U4$, $U5$ and U for $N1 = 200$.

Table 4.7 Showing values of random numbers obtained using part A of program number 4.3 for $N1 = 200$. The program performs acceptance-rejection sampling using these random numbers.

i	U1[i]	U2[i]	U3[i]	U4[i]	U5[i]	U[i]
1	4.4473	2.7149	4.4384	1.1370	4.0386	0.0192
2	2.6756	2.8983	2.5196	1.6748	1.7598	0.0015
3	2.3339	2.7028	2.5442	1.9779	3.3526	0.0097
4	3.6973	4.6357	4.5035	2.4633	2.1781	0.0280
5	4.6416	4.7104	1.2016	2.4038	1.6042	0.0073
6	2.5185	3.0958	2.4345	2.6808	3.2565	0.0057
7	1.3096	4.5698	4.9391	2.9414	2.2318	0.0206
8	4.7561	3.3692	1.7684	2.5320	3.7016	0.0123
9	4.8949	2.8002	4.1333	4.1510	3.8047	0.0247
10	3.6464	3.0806	4.6097	1.0200	1.8731	0.0098
...	...	...	...	...	...	...
191	4.9999	1.7126	1.9512	4.9061	3.3232	0.0233
192	2.7473	4.9509	1.2882	3.7747	4.8140	0.0012
193	4.4403	2.4619	3.0747	1.9924	1.8074	0.0301
194	1.7446	4.8151	1.9320	3.7212	3.7434	0.0102
195	3.8400	1.8600	2.0317	4.0704	1.1202	0.0006

196	1.5973	2.8481	3.4298	4.1709	2.0642	0.0196
197	1.4158	4.4339	2.4262	3.9055	3.2737	0.0234
198	1.1029	1.4205	1.8557	3.1847	2.2887	0.0351
199	2.3458	3.8242	3.9002	3.4212	4.2877	0.0227
200	2.2821	4.6281	4.7560	3.5151	2.7247	0.0398

Part B of program number 4.3 uses the random numbers of Table 4.7 according to acceptance-rejection sampling method and produces 5 sets of $n_1 = n_2 = n_3 = n_4 = n_5 = 11$ accepted random variates UA1, UA2, UA3, UA4 and UA5 for x_1, x_2, x_3, x_4 and x_5 respectively shown in Table 4.8 for N1 = 200. Acceptance-rejection sampling technique uses the condition $U < norc\ (U1)(U2)(1/U3)(1/U4)(1/U5)$. If the condition is satisfied, the U1, U2, U3, U4 and U5 are accepted as random variates which we call $UA1[[i]]$, $UA2[[i]]$, $UA3[[i]]$, $UA4[[i]]$ and $UA5[[i]]$ respectively for further calculations.

Table 4.8 Showing accepted values of random variates obtained using part B of program number 4.3. We have 11 accepted values out of 200 for U1, U2, U3, U4 and U5.

i	UA1[[i]]	UA2[[i]]	UA3[[i]]	UA4[[i]]	UA5[[i]]
1	2.6756	2.8983	2.5196	1.6748	1.7598
2	4.6416	4.7104	1.2016	2.4038	1.6042
3	3.2138	4.7940	1.3674	3.7497	3.3221
4	4.6375	3.4714	1.7736	4.3840	1.5407
5	4.5219	4.2191	1.7456	1.0686	1.0315
6	1.9629	4.7139	1.4850	3.7203	2.0990
7	4.3117	4.2534	3.1248	3.3145	2.2570
8	1.7357	2.5468	1.4742	1.2721	4.6477
9	3.3361	3.6937	2.1622	2.0053	1.3076
10	1.8557	4.2930	2.2207	2.9260	3.2148
11	3.8400	1.8600	2.0317	4.0704	1.1202

Using part C of program number 4.3, we evaluate the integral I_{10} using the following equation

$$I_{10} = \frac{1}{n_1\,n_2\,n_3\,n_4\,n_5} \sum_{i=1}^{n_1} \sum_{j=1}^{n_2} \sum_{k=1}^{n_3} \sum_{l=1}^{n_4} \sum_{m=1}^{n_5} \frac{F(x_{1i},x_{2j},x_{3k},x_{4l},x_{5m})}{p(x_{1i},x_{2j},x_{3k},x_{4l},x_{5m})}$$

$$= \frac{1}{n_1\,n_2\,n_3\,n_4\,n_5} \sum_{i=1}^{n_1} \sum_{j=1}^{n_2} \sum_{k=1}^{n_3} \sum_{l=1}^{n_4} \sum_{m=1}^{n_5} \frac{((x_{1i}+x_{2j})/(x_{3k}+x_{4l}+x_{5m}))^{y}}{norc\ (x_{1i})(x_{2j})(1/x_{3k})(1/x_{4l})(1/x_{5m})}$$

as

$$I_{10} = \frac{1}{n_1\,n_2\,n_3\,n_4\,n_5} \sum_{i=1}^{n_1} \sum_{j=1}^{n_2} \sum_{k=1}^{n_3} \sum_{l=1}^{n_4} \sum_{m=1}^{n_5}$$

$$\frac{(\ (\ UA1[[i]]+UA2[[j]]\ )/(UA3[[k]]+UA4[[l]]+UA5[[m]])\ )^{y}}{norc\ (\ UA1[[i]]\ )\ (\ UA2[[j]]\ )\ (\ 1/UA3[[k]]\ )\ (\ 1/UA4[[l]]\ )\ (\ 1/UA5[[m]]\ )}$$

--------(4.11)

Results are in Table 4.9 as Monte Carlo result I_{mc}.

Using part C of program number 4.3, we have also calculated the integral as I_{ni} using NIntegrate command. Results are in Table 4.9 as I_{ni}. We have compared I_{ni} with I_{mc} for various values of parameters and of N1. The difference is denoted as error1 which is % error shown in Table 4.9.

Table 4.9 Results of part C of program number 4.3 for various values of parameters. We are dealing with

$$I_{10} = \int_a^b \int_a^b \int_a^b \int_a^b \int_a^b \left(\frac{x_1 + x_2}{x_3 + x_4 + x_5} \right)^y dx_1 \ dx_2 \ dx_3 \ dx_4 \ dx_5 \ .$$

I	y	a	b	N1	n1	n1^5	Monte Carlo result I_{mc}	result obtained using NIntegrate command I_{ni}	% error
1	3	1	5	150	6	7776	591.84	535.36	10.55
2	3	1	5	180	9	59049	573.96	535.36	7.21
3	3	1	11	2000	6	7776	77323.90	79880.70	-3.20
4	2	1	5	150	6	7776	484.71	582.48	-16.79
5	2	1	5	200	11	161051	536.53	582.48	-7.89
6	1	1	5	300	14	537824	629.57	721.76	-12.77
7	1	1	5	400	18	1889568	667.05	721.76	-7.58
8	0.5	1	5	300	14	537824	692.27	844.90	-18.06
9	0.5	1	5	400	18	1889568	749.89	844.90	-11.25
10	0.5	1	5	500	19	2476099	760.31	844.90	-10.01

Table 4.9 displays a survey of Monte Carlo evaluation of the 5D definite integration I_{10} for different values of various parameters. We note that, while using Monte Carlo method, % error is generally below 10%. If the error is more than 10%, the error can usually be lowered by raising number of terms $n1^5$ in the summation in equation (4.11) by raising value of $N1$.

4.5 Evaluation of the integral I_{11}
We have evaluated the multi-dimensional definite integral

$$I_{11} = \int_a^b \int_a^b \int_g^h \int_g^h \int_g^h (x_1 + x_2 + x_3 + x_4 + x_5)^y dx_1 \ dx_2 \ dx_3 \ dx_4 \ dx_5$$

-------(4.12)

using acceptance-rejection sampling. Here $a = 1$, $b = 6$, $g = 1$, $h = 11$, $y = 0.25, 0.5, 1, 1.5, 2$.

Program number 4.4
part A

y=2

N1=N2=N3=N4=N5=200;

a=1;

b=6;

g=1;

h=11;

norc=1/(NIntegrate[x1*x2*x3*x4*x5,

{x1,a,b},{x2,a,b},{x3,g,h},{x4,g,h},{x5,g,h}])

c=norc*b*b*h*h*h

n=654321;

SeedRandom[n];

Table[{i=i+1,

U1[i]=a+(b-a)*RandomReal[],

```
U2[i]=a+(b-a)*RandomReal[],
U3[i]=g+(h-g)*RandomReal[],
U4[i]=g+(h-g)*RandomReal[],
U5[i]=g+(h-g)*RandomReal[],
U[i]=0+(c-0)*RandomReal[]},{i,0,N1-1,1}];
TableForm[%,TableSpacing->{2,2},
TableHeadings->{None,{"i","U1[i]","U2[i]","U3[i]","U4[i]","U5[i]"}}]
```

part B
```
UA1={};
i=0;
While[i<=N1,If[U[i]<norc*U1[i]*U2[i]*U3[i]*U4[i]*U5[i],
AppendTo[UA1,U1[i]]];i=i+1];
n1=Length[UA1]
```

```
UA2={};
i=0;
While[i<=N2,If[U[i]<norc*U1[i]*U2[i]*U3[i]*U4[i]*U5[i],
AppendTo[UA2,U2[i]]];i=i+1];
n2=Length[UA2]
```

```
UA3={};
i=0;
While[i<=N3,If[U[i]<norc*U1[i]*U2[i]*U3[i]*U4[i]*U5[i],
AppendTo[UA3,U3[i]]];i=i+1];
n3=Length[UA3]
```

```
UA4={};
i=0;
While[i<=N4,If[U[i]<norc*U1[i]*U2[i]*U3[i]*U4[i]*U5[i],
AppendTo[UA4,U4[i]]];i=i+1];
n4=Length[UA4]
```

```
UA5={};
i=0;
While[i<=N5,If[U[i]<norc*U1[i]*U2[i]*U3[i]*U4[i]*U5[i],
AppendTo[UA5,U5[i]]];i=i+1];
n5=Length[UA5]
```

Table[{i=i+1,UA1[[i]],UA2[[i]],UA3[[i]],UA4[[i]],UA5[[i]]},{i,0,n1-1,1}];

TableForm[%,TableSpacing->{2,2},

TableHeadings->{None,{"i","UA1[[i]]","UA2[[i]]",

"UA3[[i]]","UA4[[i]]","UA5[[i]]"}}]

part C

Imc=(1/(n1*n2*n3*n4*n5))*

Sum[(((UA1[[i]]+UA2[[j]]+UA3[[k]]+UA4[[l]]+UA5[[m]])^y)/

(norc*UA1[[i]]*UA2[[j]]*UA3[[k]]*UA4[[l]]*UA5[[m]]),

{i,1,n1},{j,1,n2},{k,1,n3},{l,1,n4},{m,1,n5}]

Ini=NIntegrate[(x1+x2+x3+x4+x5)^y,

{x1,a,b},{x2,a,b},{x3,g,h},{x4,g,h},{x5,g,h}]

error1=(Imc-Ini)*100/Ini

We have written program number 4.4 in Mathematica using symbolic computation. We have chosen a probability density function

$$p(x_1,x_2,x_3,x_4,x_5) = norc\ (x_1)(x_2)(x_3)(x_4)(x_5) \qquad \text{-------(4.13)}$$

where *norc* is normalization constant. Both the integrand $(x_1 + x_2 + x_3 + x_4 + x_5)^y$ and $p(x_1,x_2,x_3,x_4,x_5) = norc\ (x_1)(x_2)(x_3)(x_4)(x_5)$ are increasing functions of all the 5 variables x_1, x_2, x_3, x_4, x_5. Hence maximum value of p is $c = norc\ (b)(b)(h)(h)(h)$.

In part A of program number 4.4, $U1$, $U2$, $U3$, $U4$, $U5$ are 5 sets of $N1 = N2 = N3 = N4 = N5$ uniform random numbers for x_1, x_2 in the interval a to b and for x_3, x_4, x_5. In the interval g to h. U's are $N1$ uniform random numbers in the interval 0 to c.

Part B of program number 4.4 uses the random numbers U_1, U_2, U_3, U_4, U_5 according to acceptance-rejection sampling method and produces 5 sets of n_1, n_2, n_3, n_4, n_5 accepted random variates $UA1$, $UA2$, $UA3$, $UA4$, $UA5$ for x_1, x_2, x_3, x_4, x_5 shown in Table 4.10.

Acceptance-rejection sampling technique uses the condition $U < norc\ (U1)(U2)(U3)(U4)(U5)$. If the condition is satisfied, the corresponding values of $U1$, $U2$, $U3$, $U4$, $U5$ are accepted as random variates which we call $UA1[[i]]$, $UA2[[i]]$, $UA3[[i]]$, $UA4[[i]]$, $UA5[[i]]$ respectively for further calculations.

Table 4.10 Showing accepted values of random variates $U1$, $U2$, $U3$, $U4$, $U5$ obtained using part B of prog. no. 4.4. We have 11 accepted values out of 200 values of $U1$, $U2$, $U3$, $U4$, $U5$.

i	UA1[[i]]	UA2[[i]]	UA3[[i]]	UA4[[i]]	UA5[[i]]
1	2.87021	5.77819	8.29914	9.11168	10.2888
2	1.98763	4.87282	7.10596	10.9367	3.77729
3	3.76722	5.74245	1.91844	7.87428	6.80532
4	4.14318	5.67433	9.53957	10.0638	5.28478
5	2.20359	5.64239	2.21239	7.80065	3.74739
6	5.23281	3.72033	10.8549	4.22988	10.5097
7	4.39966	5.14821	7.14996	10.0156	3.69855
8	5.1396	5.06673	6.31193	6.78615	4.14256

9	4.39555	2.68264	9.69107	2.73888	10.0023
10	2.06961	5.11627	4.05177	5.81494	6.53696
11	3.18409	5.93857	1.72048	7.93676	10.535

Using part C of program number 4.4, we evaluate the integral I_{11} using the following equation

$$I_{11} = \frac{1}{n_1\, n_2\, n_3\, n_4\, n_5} \sum_{i=1}^{n_1} \sum_{j=1}^{n_2} \sum_{k=1}^{n_3} \sum_{l=1}^{n_4} \sum_{m=1}^{n_5} \frac{F(x_{1i}, x_{2j}, x_{3k}, x_{4l}, x_{5m})}{p(x_{1i}, x_{2j}, x_{3k}, x_{4l}, x_{5m})}$$

$$= \frac{1}{n_1\, n_2\, n_3\, n_4\, n_5} \sum_{i=1}^{n_1} \sum_{j=1}^{n_2} \sum_{k=1}^{n_3} \sum_{l=1}^{n_4} \sum_{m=1}^{n_5} \frac{(x_{1i} + x_{2j} + x_{3k} + x_{4l} + x_{5m})^y}{norc\ (x_{1i})(x_{2j})(x_{3k})(x_{4l})(x_{5m})}$$

as

$$I_{11} = \frac{1}{n_1\, n_2\, n_3\, n_4\, n_5} \sum_{i=1}^{n_1} \sum_{j=1}^{n_2} \sum_{k=1}^{n_3} \sum_{l=1}^{n_4} \sum_{m=1}^{n_5}$$

$$\frac{(\ UA1[[i]] + UA2[[j]] + UA3[[k]] + UA4[[l]] + UA5[[m]])^y}{norc\ (UA1[[i]])\ (UA2[[j]])\ (UA3[[k]])\ (UA4[[l]])\ (UA5[[m]])}$$

--------(4.14)

Results are in Table 4.11 as Monte Carlo results I_{mc}.

Using part C of program number 4.4, we have also calculated the integral as I_{ni} using NIntegrate command. Results are in Table 4.11 as I_{ni}. We have compared I_{ni} with I_{mc} for various values of parameters. The difference has been shown as % error in Table 4.11. Results obtained using program number 4.4 are as shown in Table 4.11.

Table 4.11 Results obtained using program number 4.4. Using $a = 1$, $b = 6$, $g = 1$, $h = 11$ for various values of parameters. We are dealing with

$$I_{11} = \int_a^b \int_a^b \int_g^h \int_g^h \int_g^h (x_1 + x_2 + x_3 + x_4 + x_5)^y\, dx_1\, dx_2\, dx_3\, dx_4\, dx_5$$

y	$N1$	$n1$	$(n1)^5$	Monte Carlo result I_{mc}	results obtained using NIntegrate command I_{ni}	% error
2	200	11	161051	1.74496*10^7	1.63542*10^7	6.70
2	150	7	16807	1.54522*10^7	1.63542*10^7	-5.51
2	100	2	32	1.21086*10^7	1.63542*10^7	-25.96
1.5	100	2	32	2.16903*10^6	3.18014*10^6	-31.8
1.5	150	7	16807	2.97751*10^6	3.18014*10^6	-6.37
1.5	200	11	161051	3.45085*10^6	3.18014*10^6	8.51
1.5	250	15	759375	3.41488*10^6	3.18014*10^6	7.38
1	250	15	759375	679576	625000.	8.73
1	200	11	161051	689326.	625000.	10.3
1	150	7	16807	578542.	625000.	-7.43
1	100	2	32	389581.	625000.	-37.7
0.5	100	2	32	70155.6	124239.	-43.5
0.5	150	7	16807	113374.	124239.	-8.75
0.5	200	11	161051	139126.	124239.	11.98
0.5	250	15	759375	136737.	124239.	10.06
0.25	100	2	32	29799.5	55643.1	-46.4
0.25	150	7	16807	50351.2	55643.1	-9.5
0.25	200	11	161051	62750.8	55643.1	12.77
0.25	250	15	759375	61597.4	55643.1	10.7

4.6 Evaluation of the integral I_{12}

We have evaluated the multi-dimensional definite integral

$$I_{12} = \int_a^b \int_a^b \int_p^q \int_p^q \int_p^q \left(\frac{x_1 + x_2}{x_3 + x_4 + x_5} \right)^{-y} dx_1\ dx_2\ dx_3\ dx_4\ dx_5 \qquad \text{--------(4.15)}$$

using acceptance-rejection sampling. Here we take $a = 1$, $b = 2$, $p = 1$, $q = 3$, $y = 1, 1.5, 2, 2.5, 3$. We have written program number 4.5 in Mathematica using symbolic computation.

Program number 4.5
part A

```
y=3

N1=N2=N3=N4=N5=100;

a=1;

b=2;

p=1;

q=3;

norc=1/(NIntegrate[(1/x1)*(1/x2)*(x3)*(x4)*(x5),

{x1,a,b},{x2,a,b},{x3,p,q},{x4,p,q},{x5,p,q}])

c=norc*(1/a)*(1/a)*(q)*(q)*(q)

n=654321;

SeedRandom[n];

Table[{i=i+1,

U1[i]=a+(b-a)*RandomReal[],

U2[i]=a+(b-a)*RandomReal[],

U3[i]=p+(q-p)*RandomReal[],

U4[i]=p+(q-p)*RandomReal[],

U5[i]=p+(q-p)*RandomReal[],

U[i]=0+(c-0)*RandomReal[]},{i,0,N1-1,1}];

TableForm[%,TableSpacing->{2,2},

TableHeadings->{None,{"i","U1[i]","U2[i]","U3[i]",

"U4[i]","U5[i]","U[i]"}}]
```

part B

```
UA1={};

i=0;

While[i<=N1,

If[U[i]<norc*(1/U1[i])*(1/U2[i])*(U3[i])*(U4[i])*(U5[i]),

AppendTo[UA1,U1[i]]];i=i+1];

n1=Length[UA1]
```

```
UA2={};
i=0;
While[i<=N2,
If[U[i]<norc*(1/U1[i])*(1/U2[i])*(U3[i])*(U4[i])*(U5[i]),
AppendTo[UA2,U2[i]]];i=i+1];
n2=Length[UA2]

UA3={};
i=0;
While[i<=N3,
If[U[i]<norc*(1/U1[i])*(1/U2[i])*(U3[i])*(U4[i])*(U5[i]),
AppendTo[UA3,U3[i]]];i=i+1];
n3=Length[UA3]

UA4={};
i=0;
While[i<=N4,
If[U[i]<norc*(1/U1[i])*(1/U2[i])*(U3[i])*(U4[i])*(U5[i]),
AppendTo[UA4,U4[i]]];i=i+1];
n4=Length[UA4]

UA5={};
i=0;
While[i<=N5,
If[U[i]<norc*(1/U1[i])*(1/U2[i])*(U3[i])*(U4[i])*(U5[i]),
AppendTo[UA5,U5[i]]];i=i+1];
n5=Length[UA5]


Table[{i=i+1,UA1[[i]],UA2[[i]],UA3[[i]],UA4[[i]],UA5[[i]]},{i,0,n1-1,1}];
TableForm[%,TableSpacing->{2,2},
TableHeadings->{None,{"i","UA1[[i]]","UA2[[i]]",
"UA3[[i]]","UA4[[i]]","UA5[[i]]"}}]
```

part C

```
Imc=(1/(n1*n2*n3*n4*n5))*
Sum[((((UA1[[i]]+UA2[[j]])/(UA3[[k]]+UA4[[l]]+UA5[[m]]))^(-y))/
(norc*(1/UA1[[i]])*(1/UA2[[j]])*(UA3[[k]])*(UA4[[l]])*(UA5[[m]])),
{i,1,n1},{j,1,n2},{k,1,n3},{l,1,n4},{m,1,n5}]
```

Ini=NIntegrate[((x1+x2)/(x3+x4+x5))^(-y),

{x1,a,b},{x2,a,b},{x3,p,q},{x4,p,q},{x5,p,q}]

error1=(Imc-Ini)*100/Ini

We have chosen a probability density function

$$p(x_1, x_2, x_3, x_4, x_5) = norc \ (1/x_1) \ (1/x_2) \ (x_3) \ (x_4) \ (x_5) \qquad\qquad ------(4.16)$$

where   *norc*   is   normalization   constant.   Both   the   integrand $\left(\dfrac{x_1 + x_2}{x_3 + x_4 + x_5} \right)^{-y}$ and

$p(x_1, x_2, x_3, x_4, x_5) = norc \ (1/x_1) \ (1/x_2) \ (x_3) \ (x_4) \ (x_5)$ are decreasing functions of x_1 and x_2 and increasing

function of x_3, x_4 and x_5. Hence maximum value of p is $c = norc(1/a)(1/a)(q)(q)(q)$.

In part A of program number 4.5, U1, U2, U3, U4 and U5 are 5 sets of $N1 = N2 = N3 = N4 = N5$ uniform random numbers for x_1, x_2, x_3, x_4, x_5 respectively in the interval a to b for U1 and U2 and in the interval p to q for U3, U4 and U5. U's are N_1 uniform random numbers in the interval 0 to c.

Part B of program number 4.5 uses the random numbers U_1, U_2, U_3, U_4, U_5 and U according to acceptance-rejection sampling method and produces 5 sets of $n_1 = n_2 = n_3 = n_4 = n_5$ accepted random variates UA1, UA2, UA3, UA4 and UA5 for x_1, x_2, x_3, x_4, x_5 respectively shown in Table 4.12. Acceptance-rejection sampling technique uses the condition $U < norc \ (1/U1)(1/U2)(U3)(U4)(U5)$. If the condition is satisfied, the corresponding values of U1, U2, U3, U4, U5 are accepted as random variates which we call UA1[[i]], UA2[[i]], UA3[[i]] , UA4[[i]], UA5[[i]] respectively for further calculations.

Table 4.12 Showing accepted values of random variates U1, U2, U3, U4, U5 obtained using part B of prog. no. 4.5. We have 13 accepted values out of 100 values of U1, U2, U3, U4, U5.

i	UA1[[i]]	UA2[[i]]	UA3[[i]]	UA4[[i]]	UA5[[i]]
1	1.41891	1.47457	1.75982	1.33739	1.37988
2	1.1655	1.12183	1.70877	1.70463	1.77894
3	1.32606	1.19211	2.85874	1.19413	2.73567
4	1.27896	1.30776	2.86716	2.43454	2.92894
5	1.1165	1.66134	2.84975	2.26215	2.25519
6	1.32613	1.32252	2.46992	1.97704	1.9562
7	1.37404	1.95564	2.45983	2.62234	2.85775
8	1.22249	1.02699	2.72056	1.74682	2.37039
9	1.91107	1.32495	2.21976	1.5209	1.44673
10	1.19753	1.77456	2.22119	2.98735	1.55546
11	1.69282	1.14996	2.31145	2.09723	2.38711
12	1.18051	1.56239	1.5413	1.64986	2.3208
13	1.02308	1.04819	2.54065	2.78331	1.9815

Using part C of prog no. 4.5, we evaluate the integral I_{12} using the following equation

$$I_{12} = \frac{1}{n_1\, n_2\, n_3\, n_4\, n_5} \sum_{i=1}^{n_1} \sum_{j=1}^{n_2} \sum_{k=1}^{n_3} \sum_{l=1}^{n_4} \sum_{m=1}^{n_5} \frac{F(x_{1i}, x_{2j}, x_{3k}, x_{4l}, x_{5m})}{p(x_{1i}, x_{2j}, x_{3k}, x_{4l}, x_{5m})}$$

$$= \frac{1}{n_1\, n_2\, n_3\, n_4\, n_5} \sum_{i=1}^{n_1} \sum_{j=1}^{n_2} \sum_{k=1}^{n_3} \sum_{l=1}^{n_4} \sum_{m=1}^{n_5} \frac{((x_{1i} + x_{2j})/(x_{3k} + x_{4l} + x_{5m}))^{-y}}{norc \ (1/x_{1i})(1/x_{2j})(x_{3k})(x_{4l})(x_{5m})}$$

as

$$I_{12} = \frac{1}{n_1 \, n_2 \, n_3 \, n_4 \, n_5} \sum_{i=1}^{n_1} \sum_{j=1}^{n_2} \sum_{k=1}^{n_3} \sum_{l=1}^{n_4} \sum_{m=1}^{n_5}$$

------(4.17)

$$\frac{\left(\, (\, UA1[[i]] + UA2[[j]] \,) \, /(UA3[[k]] + UA4[[l]] + UA5[[m]]) \, \right)^{-y}}{norc \, (\, 1/UA1[[i]] \,) \, (\, 1/UA2[[j]] \,) \, (\, UA3[[k]] \,) \, (\, UA4[[l]] \,) \, (\, UA5[[m]] \,)}$$

Results are in Table 4.13 as Monte Carlo results I_{mc}.

Using part C of program number 4.5, we have also calculated the integral as I_{ni} using NIntegrate command. Results are in Table 4.13 as I_{ni}. We have compared I_{ni} with I_{mc} for various values of parameters. The difference has been shown as % error in Table 4.13. Results obtained using program number 4.5 are as shown in Table 4.13.

Table 4.13 Results obtained using program number 4.5 using various values of parameters. We are dealing with

$$I_{12} = \int_a^b \int_a^b \int_p^q \int_p^q \int_p^q \left(\frac{x_1 + x_2}{x_3 + x_4 + x_5} \right)^{-y} dx_1 \, dx_2 \, dx_3 \, dx_4 \, dx_5 \text{ using } a = 1, b = 2, p = 1, q = 3$$

y	$N1=$ $N2=$ $N3=$ $N4=$ $N5$	$n1=$ $n2=$ $n3=$ $n4= n5$	$(n1)^5$	Monte Carlo result I_{mc}	result obtained using NIntegrate command I_{ni}	% error
3	100	13	371293	82.0446	78	5.19
3	50	8	32768	84.1172	78	7.84
3	150	21	4084101	77.5774	78	-0.54
3	25	5	3125	87.5193	78	12.20
2.5	25	5	3125	56.1035	51.857	8.19
2.5	50	8	32768	53.1574	51.857	2.51
2.5	100	13	371293	52.6243	51.857	1.48
2.5	150	21	4084101	52.0205	51.857	0.32
2	150	21	4084101	35.2475	34.8638	1.10
2	100	13	371293	34.0479	34.8638	-2.34
2	50	8	32768	33.8346	34.8638	-2.95
2	25	5	3125	36.2321	34.8638	3.92
1.5	25	5	3125	23.5772	23.708	-0.55
1.5	50	8	32768	21.6954	23.708	-8.49
1.5	100	13	371293	22.2252	23.708	-6.25
1.5	150	21	4084101	24.1361	23.708	1.81
1	150	21	4084101	16.7053	16.3103	2.42
1	100	13	371293	14.6395	16.3103	-10.24
1	50	8	32768	14.0174	16.3103	-14.06
1	25	5	3125	15.4615	16.3103	-5.20
3.5	100	13	371293	129.002	118.613	8.76
3.5	150	21	4084101	116.878	118.613	-1.46
3.5	50	8	32768	134.04	118.613	13.01
3.5	25	5	3125	137.517	118.613	15.94

Chapter V

Evaluation of Seven-dimensional Definite Integrals

.

This chapter deals with Monte Carlo evaluation of 2 seven-dimensional definite integrals. In each case, a suitable multi-variable probability density function is chosen and acceptance-rejection sampling is used to obtain values of 7 sets of random variates corresponding to the 7 random variables in the probability density function. The integrals are evaluated using the 7 sets of random variates. Programs written in Mathematica have been used in the sampling as well as in evaluating the integrals. Uses of different parts of the programs have been narrated.

5.1 The 7D definite integrals dealt with in this chapter
This chapter deals with the following 2 definite integrals:

$$I_{13} = \int_a^b \int_a^b \int_a^b \int_a^b \int_a^b \int_a^b \int_a^b (x_1 + x_2 + x_3 + x_4 + x_5 + x_6 + x_7)^y \, dx_1 \, dx_2 \, dx_3 \, dx_4 \, dx_5 \, dx_6 \, dx_7$$

$$--------(5.1)$$

$$I_{14} = \int_a^b \int_a^b \int_a^b \int_a^b \int_a^b \int_a^b \int_a^b \left(\frac{x_1 + x_2 + x_3}{x_4 + x_5 + x_6 + x_7} \right)^y \, dx_1 \, dx_2 \, dx_3 \, dx_4 \, dx_5 \, dx_6 \, dx_7$$

$$--------(5.2)$$

where y is positive constant such as 2, 3 etc.; limits a and b are also constants such as 1, 3, 5 etc.

5.2 Evaluation of the integral I_{13}
We now deal with the definite integral I_{13}.

$$I_{13} = \int_a^b \int_a^b \int_a^b \int_a^b \int_a^b \int_a^b \int_a^b (x_1 + x_2 + x_3 + x_4 + x_5 + x_6 + x_7)^y \, dx_1 \, dx_2 \, dx_3 \, dx_4 \, dx_5 \, dx_6 \, dx_7$$

We first choose a probability density function

$$p(x_1, x_2, x_3, x_4, x_5, x_6, x_7) = norc \ (x_1) \ (x_2) \ (x_3) \ (x_4) \ (x_5) \ (x_6) \ (x_7) \qquad ----(5.3)$$

where *norc* is normalization constant. Both $F(x_1, x_2, x_3, x_4, x_5, x_6, x_7) = (x_1 + x_2 + x_3 + x_4 + x_5 + x_6 + x_7)^y$ and $p(x_1, x_2, x_3, x_4, x_5, x_6, x_7) = norc \ (x_1) \ (x_2) \ (x_3) \ (x_4) \ (x_5) \ (x_6) \ (x_7)$ are increasing functions of x_1, x_2, x_3, x_4, x_5, x_6 and x_7. The probability density function can be normalized without running into the problem of evaluating another multi-dimensional definite integral. Maximum value of the probability density function is $c = norc$ $(b)(b)(b)(b)(b)(b)(b)$.

Program number 5.1
part A

y=3

N1=200;

a=1;

b=3;

norc=1/(NIntegrate[(x1)*(x2)*(x3)*(x4)*(x5)*(x6)*(x7),

{x1,a,b},{x2,a,b},{x3,a,b},{x4,a,b},{x5,a,b},{x6,a,b},{x7,a,b}])

c=norc*(b)*(b)*(b)*(b)*(b)*(b)*(b)

n=654321;

SeedRandom[n];

Table[{i=i+1,

U1[i]=a+(b-a)*RandomReal[],

```
U2[i]=a+(b-a)*RandomReal[],

U3[i]=a+(b-a)*RandomReal[],

U4[i]=a+(b-a)*RandomReal[],

U5[i]=a+(b-a)*RandomReal[],

U6[i]=a+(b-a)*RandomReal[],

U7[i]=a+(b-a)*RandomReal[],

u[i]=RandomReal[];U[i]=0+(c-0)*u[i]},{i,0,N1-1,1}];

TableForm[%,TableSpacing->{2,2},

TableHeadings->{None,{"i","U1[i]","U2[i]","U3[i]",

"U4[i]","U5[i]","U6[i]","U7[i]","U[i]"}}]
```

part B

```
UA1={};

i=0;

While[i<=N1,If[U[i]<norc*U1[i]*U2[i]*U3[i]*U4[i]*U5[i]*U6[i]*U7[i],

AppendTo[UA1,U1[i]]];i=i+1];

n1=Length[UA1]


UA2={};

i=0;

While[i<=N1,If[U[i]<norc*U1[i]*U2[i]*U3[i]*U4[i]*U5[i]*U6[i]*U7[i],

AppendTo[UA2,U2[i]]];i=i+1];

n2=Length[UA2]


UA3={};

i=0;

While[i<=N1,If[U[i]<norc*U1[i]*U2[i]*U3[i]*U4[i]*U5[i]*U6[i]*U7[i],

AppendTo[UA3,U3[i]]];i=i+1];

n3=Length[UA3]


UA4={};

i=0;

While[i<=N1,If[U[i]<norc*U1[i]*U2[i]*U3[i]*U4[i]*U5[i]*U6[i]*U7[i],

AppendTo[UA4,U4[i]]];i=i+1];

n4=Length[UA4]


UA5={};

i=0;
```

While[i<=N1,If[U[i]<norc*U1[i]*U2[i]*U3[i]*U4[i]*U5[i]*U6[i]*U7[i],

AppendTo[UA5,U5[i]]];i=i+1];

n5=Length[UA5]

UA6={};

i=0;

While[i<=N1,If[U[i]<norc*U1[i]*U2[i]*U3[i]*U4[i]*U5[i]*U6[i]*U7[i],

AppendTo[UA6,U6[i]]];i=i+1];

n6=Length[UA6]

UA7={};

i=0;

While[i<=N1,If[U[i]<norc*U1[i]*U2[i]*U3[i]*U4[i]*U5[i]*U6[i]*U7[i],

AppendTo[UA7,U7[i]]];i=i+1];

n7=Length[UA7]

Table[{i=i+1,UA1[[i]],UA2[[i]],UA3[[i]],UA4[[i]],

UA5[[i]],UA6[[i]],UA7[[i]]},{i,0,n1-1,1}];

TableForm[%,TableSpacing->{2,2},

TableHeadings->{None,{"i","UA1[[i]]","UA2[[i]]","UA3[[i]]",

"UA4[[i]]","UA5[[i]]","UA6[[i]]","UA7[[i]]"}}]

part C

Imc=(1/(n1*n2*n3*n4*n5*n6*n7))*

Sum[(((UA1[[i]]+UA2[[j]]+UA3[[k]]+UA4[[l]]+UA5[[m]]+UA6[[n]]+UA7[[o]])^y)/

(norc*UA1[[i]]*UA2[[j]]*UA3[[k]]*UA4[[l]]*UA5[[m]]*UA6[[n]]*UA7[[o]]),

{i,1,n1},{j,1,n2},{k,1,n3},{l,1,n4},{m,1,n5},{n,1,n6},{o,1,n7}]

Ini=NIntegrate[(x1+x2+x3+x4+x5+x6+x7)^y,

{x1,a,b},{x2,a,b},{x3,a,b},{x4,a,b},{x5,a,b},{x6,a,b},{x7,a,b}]

error1=(Imc-Ini)*100/Ini

To evaluate the 7D definite integral I_{13}, we have written program number 5.1. In part A of program number 5.1, $U1$, $U2$, $U3$, $U4$, $U5$, $U6$ and $U7$ are 7 sets of $N1$ uniform random numbers for x_1, x_2, x_3, x_4, x_5, x_6, x_7 respectively in the interval a to b. u and U are 2 sets of $N1$ uniform random numbers in interval 0 to 1 and 0 to c respectively. Table 5.1 shows these random numbers $U1$, $U2$, $U3$, $U4$, $U5$, $U6$, $U7$ and U for $N1 = 200$.

Table 5.1 Showing values of random numbers obtained using part A of program number 5.1 for $N1$ = 200. The program performs acceptance-rejection sampling using these random numbers.

i	U1[i]	U2[i]	U3[i]	U4[i]	U5[i]	U6[i]	U7[i]	U[i]
1	2.7237	1.8574	2.7192	1.0685	2.5193	1.9217	1.8378	0.0633
2	1.7598	1.3374	1.3799	1.0729	1.6669	1.8514	1.7721	0.0326
3	2.1763	1.4644	2.3486	2.8179	2.7518	1.7316	1.5891	0.0897
4	2.8208	2.8552	1.1008	1.7019	1.3021	1.3523	1.7593	0.0699
5	1.7172	1.8404	2.1283	1.2743	1.1548	2.7849	2.9696	0.0648
6	1.6159	1.9890	2.8781	2.1846	1.3842	1.7660	2.3508	0.0393
7	2.9474	1.9001	2.5667	2.5755	2.4024	2.1875	2.3232	0.0694
8	2.8049	1.0100	1.4365	1.4722	1.8954	2.4615	1.6857	0.0302
9	1.2158	1.9758	1.3310	1.2437	1.7088	1.7046	1.7789	0.0122
10	2.9262	2.0368	2.2656	2.6516	1.8201	2.3484	1.4176	0.1152
...	...	...	...	...	...	...	...	...
191	2.5356	1.9685	1.0684	2.8125	1.8110	2.3171	1.6751	0.0918
192	1.6621	1.6621	1.1061	2.4615	1.2742	2.9620	2.5520	0.1131
193	2.1081	2.9655	2.8327	1.4297	1.6025	1.0156	2.0492	0.1224
194	1.9332	1.7305	1.0552	1.5202	1.0407	2.5421	2.9429	0.0807
195	2.1030	2.7523	1.8612	1.4648	1.4618	1.6822	2.9735	0.0418
196	2.5614	1.1869	2.2864	1.2580	2.3202	2.4341	1.5767	0.0829
197	2.1673	1.3503	2.9004	1.4193	2.9567	1.7438	2.2231	0.1093
198	1.6739	1.0973	1.9371	1.7481	2.2696	2.0331	1.4356	0.1105
199	1.8737	2.3640	2.4110	1.8138	1.2346	1.2542	1.4041	0.1271
200	2.5318	2.2498	1.5426	2.6429	1.2606	1.2351	1.2865	0.0789

Part B of program number 5.1 uses the random numbers of Table 5.1 according to acceptance-rejection sampling method and produces 7 sets of $n_1 = n_2 = n_3 = n_4 = n_5 = n_6 = n_7 = 9$ accepted random variates UA1, UA2, UA3, UA4, UA5, UA6 and UA7 for x_1, x_2, x_3, x_4, x_5, x_6, x_7 respectively shown in Table 5.2 for $N1$ = 200. Acceptance-rejection sampling technique uses the condition $U < norc\ (U1)(U2)(U3)(U4)(U5)(U6)(U7)$. If the condition is satisfied, the $U1$, $U2$, $U3$, $U4$, $U5$, $U6$ and $U7$ are accepted as random variates which we call $UA1[[i]]$, $UA2[[i]]$, $UA3[[i]]$, $UA4[[i]]$, $UA5[[i]]$, $UA6[[i]]$ and $UA7[[i]]$ respectively for further calculations.

Table 5.2 Showing accepted values of random variates obtained using part B of program number 5.1. We have 9 accepted values out of 200 for $U1$, $U2$, $U3$, $U4$, $U5$, $U6$ and $U7$.

i	UA1[[i]]	UA2[[i]]	UA3[[i]]	UA4[[i]]	UA5[[i]]	UA6[[i]]	UA7[[i]]
1	1.4084	1.6054	1.1007	2.6925	1.8267	1.4903	2.1130
2	2.0337	1.7139	2.8156	1.3348	2.4258	1.7191	2.7864
3	2.2163	2.3881	2.7895	2.3139	2.4195	2.5045	2.8608
4	1.0220	2.7222	2.1069	2.8970	1.1837	2.3749	2.1611
5	2.1373	2.8294	2.1605	2.7108	1.5260	2.5290	1.1250
6	2.1961	1.5937	1.8595	1.3074	2.0610	1.3927	2.3411
7	2.4071	1.7721	2.6558	2.6267	2.0624	2.1572	1.6285
8	1.6840	2.0032	2.2000	2.5506	2.7777	1.5358	2.5734
9	2.1616	2.1209	1.8736	2.9754	1.1441	2.3874	2.9070

Using part C of program number 5.1, we evaluate the integral I_{13} using the following equation

$$I_{13} = \frac{1}{n_1\,n_2\,n_3\,n_4\,n_5\,n_6\,n_7} \sum_{i=1}^{n_1} \sum_{j=1}^{n_2} \sum_{k=1}^{n_3} \sum_{l=1}^{n_4} \sum_{m=1}^{n_5} \sum_{n=1}^{n_6} \sum_{o=1}^{n_7}$$

$$\frac{F(x_{1i}, x_{2j}, x_{3k}, x_{4l}, x_{5m}, x_{6n}, x_{7o})}{p(x_{1i}, x_{2j}, x_{3k}, x_{4l}, x_{5m}, x_{6n}, x_{7o})}$$

$$= \frac{1}{n_1\,n_2\,n_3\,n_4\,n_5\,n_6\,n_7} \sum_{i=1}^{n_1} \sum_{j=1}^{n_2} \sum_{k=1}^{n_3} \sum_{l=1}^{n_4} \sum_{m=1}^{n_5} \sum_{n=1}^{n_6} \sum_{o=1}^{n_7}$$

$$\frac{(x_{1i} + x_{2j} + x_{3k} + x_{4l} + x_{5m} + x_{6n} + x_{7o})^y}{norc\ (x_{1i})(x_{2j})(x_{3k})(x_{4l})(x_{5m})(x_{6n})(x_{7o})}$$

as

$$I_{13} = \frac{1}{n_1\,n_2\,n_3\,n_4\,n_5\,n_6\,n_7} \sum_{i=1}^{n_1} \sum_{j=1}^{n_2} \sum_{k=1}^{n_3} \sum_{l=1}^{n_4} \sum_{m=1}^{n_5} \sum_{n=1}^{n_6} \sum_{o=1}^{n_7}$$

$$\frac{(\ UA1[[i]] + UA2[[j]] + UA3[[k]] + UA4[[l]] + UA5[[m]] + UA6[[n]] + UA7[[o]])^y}{norc\ (UA1[[i]])\ (UA2[[j]])\ (UA3[[k]])\ (UA4[[l]])\ (UA5[[m]])\ (UA6[[n]])\ (UA7[[o]])}$$

--------(5.4)

Results are in Table 5.3 as Monte Carlo result I_{mc}.

Using part C of program number 5.1, we have also calculated the integral as I_{ni} using NIntegrate command. Results are in Table 5.3 as I_{ni}. We have compared I_{ni} with I_{mc} for various values of parameters and of $N1$. The difference is denoted as error1 which is % error shown in Table 5.3.

Table 5.3 Results of part C of program number 5.1 for various values of parameters. We are dealing with

$$I_{13} = \int_a^b \int_a^b \int_a^b \int_a^b \int_a^b \int_a^b \int_a^b (x_1 + x_2 + x_3 + x_4 + x_5 + x_6 + x_7)^y\, dx_1\ dx_2\ dx_3\ dx_4\ dx_5\ dx_6\ dx_7$$

i	y	a	b	$N1$	$n1$	$n1^7$	Monte Carlo result I_{mc}	result obtained using NIntegrate command I_{ni}	% error
1	3	1	3	200	9	4782969	396988.00	363776.00	9.13
2	3	1	5	500	9	4782969	1.75*10^8	1.61*10^8	8.60
3	2	1	3	200	9	4782969	28276.40	25386.70	11.38
4	2	1	3	250	11	19487171	26810.30	25386.70	5.61
5	2	1	5	500	9	4782969	8.06*10^6	7.38*10^6	9.26
6	2	1	5	700	11	19487171	7.36*10^6	7.38*10^6	-0.18

Table 5.3 displays a survey of Monte Carlo evaluation of the 7D definite integration I_{13} for different values of various parameters. We note that, while using Monte Carlo method, % error is generally below 10%. If the error is more than 10%, the error can usually be lowered by raising number of terms $n1^7$ in the summation in equation (5.4) by raising value of $N1$.

5.3 Evaluation of the integral I_{14}

We now deal with the definite integral I_{14}.

$$I_{14} = \int_a^b \int_a^b \int_a^b \int_a^b \int_a^b \int_a^b \int_a^b \left(\frac{x_1 + x_2 + x_3}{x_4 + x_5 + x_6 + x_7} \right)^y dx_1 \; dx_2 \; dx_3 \; dx_4 \; dx_5 \; dx_6 \; dx_7$$

where y is positive. We first choose a probability density function

$$p(x_1, x_2, x_3, x_4, x_5, x_6, x_7) = norc \; (x_1) \; (x_2) \; (x_3) \; (1/x_4) \; (1/x_5) \; (1/x_6) \; (1/x_7) \qquad\qquad ----(5.5)$$

where *norc* is normalization constant. Both $F(x_1, x_2, x_3, x_4, x_5, x_6, x_7) = \left(\dfrac{x_1 + x_2 + x_3}{x_4 + x_5 + x_6 + x_7} \right)^y$ and

$p(x_1, x_2, x_3, x_4, x_5, x_6, x_7) = norc \; (x_1) \; (x_2) \; (x_3) \; (1/x_4) \; (1/x_5) \; (1/x_6) \; (1/x_7)$ are increasing functions

of x_1, x_2, x_3 and decreasing functions of x_4, x_5, x_6 and x_7. The probability density function can be normalized without running into the problem of evaluating another multi-dimensional definite integral. Maximum value of the probability density function is $c = norc \; (b)(b)(b)(1/a)(1/a)(1/a)(1/a)$.

Program number 5.2
part A

```
y=1
N1=500;
a=1;
b=3;
norc=1/(NIntegrate[(x1)*(x2)*(x3)*(1/x4)*(1/x5)*(1/x6)*(1/x7),
{x1,a,b},{x2,a,b},{x3,a,b},{x4,a,b},{x5,a,b},{x6,a,b},{x7,a,b}])
c=norc*(b)*(b)*(b)*(1/a)*(1/a)*(1/a)*(1/a)
n=654321;
SeedRandom[n];
Table[{i=i+1,
U1[i]=a+(b-a)*RandomReal[],
U2[i]=a+(b-a)*RandomReal[],
U3[i]=a+(b-a)*RandomReal[],
U4[i]=a+(b-a)*RandomReal[],
U5[i]=a+(b-a)*RandomReal[],
U6[i]=a+(b-a)*RandomReal[],
U7[i]=a+(b-a)*RandomReal[],
u[i]=RandomReal[];U[i]=0+(c-0)*u[i]},{i,0,N1-1,1}];
TableForm[%,TableSpacing->{2,2},
TableHeadings->{None,{"i","U1[i]","U2[i]","U3[i]",
"U4[i]", "U5[i]","U6[i]","U7[i]","U[i]"}}]
```

part B
```
UA1={};
```

```
i=0;
While[i<=N1,
If[U[i]<norc*U1[i]*U2[i]*U3[i]*(1/U4[i])*(1/U5[i])*(1/U6[i])*(1/U7[i]),
AppendTo[UA1,U1[i]]];i=i+1];
n1=Length[UA1]

UA2={};
i=0;
While[i<=N1,
If[U[i]<norc*U1[i]*U2[i]*U3[i]*(1/U4[i])*(1/U5[i])*(1/U6[i])*(1/U7[i]),
AppendTo[UA2,U2[i]]];i=i+1];
n2=Length[UA2]

UA3={};
i=0;
While[i<=N1,
If[U[i]<norc*U1[i]*U2[i]*U3[i]*(1/U4[i])*(1/U5[i])*(1/U6[i])*(1/U7[i]),
AppendTo[UA3,U3[i]]];i=i+1];
n3=Length[UA3]

UA4={};
i=0;
While[i<=N1,
If[U[i]<norc*U1[i]*U2[i]*U3[i]*(1/U4[i])*(1/U5[i])*(1/U6[i])*(1/U7[i]),
AppendTo[UA4,U4[i]]];i=i+1];
n4=Length[UA4]

UA5={};
i=0;
While[i<=N1,
If[U[i]<norc*U1[i]*U2[i]*U3[i]*(1/U4[i])*(1/U5[i])*(1/U6[i])*(1/U7[i]),
AppendTo[UA5,U5[i]]];i=i+1];
n5=Length[UA5]

UA6={};
i=0;
While[i<=N1,
If[U[i]<norc*U1[i]*U2[i]*U3[i]*(1/U4[i])*(1/U5[i])*(1/U6[i])*(1/U7[i]),
```

```
AppendTo[UA6,U6[i]]];i=i+1];
n6=Length[UA6]


UA7={};
i=0;
While[i<=N1,
If[U[i]<norc*U1[i]*U2[i]*U3[i]*(1/U4[i])*(1/U5[i])*(1/U6[i])*(1/U7[i]),
AppendTo[UA7,U7[i]]];i=i+1];
n7=Length[UA7]


Table[{i=i+1,UA1[[i]],UA2[[i]],UA3[[i]],UA4[[i]],UA5[[i]],UA6[[i]],UA7[[i]]},
{i,0,n1-1,1}];
TableForm[%,TableSpacing->{2,2},
TableHeadings->{None,{"i","UA1[[i]]","UA2[[i]]","UA3[[i]]",
"UA4[[i]]","UA5[[i]]","UA6[[i]]","UA7[[i]]"}}]
```

part C

```
Imc=(1/(n1*n2*n3*n4*n5*n6*n7))*
Sum[((((UA1[[i]]+UA2[[j]]+UA3[[k]])/(UA4[[l]]+UA5[[m]]+UA6[[n]]+UA7[[o]]))^y)/
(norc*UA1[[i]]*UA2[[j]]*UA3[[k]]*(1/UA4[[l]])*
(1/UA5[[m]])*(1/UA6[[n]])*(1/UA7[[o]])),
{i,1,n1},{j,1,n2},{k,1,n3},{l,1,n4},{m,1,n5},{n,1,n6},{o,1,n7}]


Ini=NIntegrate[((x1+x2+x3)/(x4+x5+x6+x7))^y,
{x1,a,b},{x2,a,b},{x3,a,b},{x4,a,b},{x5,a,b},{x6,a,b},{x7,a,b}]
error1=(Imc-Ini)*100/Ini
```

To evaluate the 7D definite integral I_{14}, we have written program number 5.2. In part A of program number 5.2, $U1$, $U2$, $U3$, $U4$, $U5$, $U6$ and $U7$ are 7 sets of $N1$ uniform random numbers for x_1, x_2, x_3, x_4, x_5, x_6, x_7 respectively in the interval a to b. u and U are 2 sets of $N1$ uniform random numbers in interval 0 to 1 and 0 to c respectively. Table 5.4 shows these random numbers $U1$, $U2$, $U3$, $U4$, $U5$, $U6$, $U7$ and U for $N1$ = 500.

Table 5.4 Showing values of random numbers obtained using part A of program number 5.2 for $N1$ = 500. The program performs acceptance-rejection sampling using these random numbers.

i	U1[i]	U2[i]	U3[i]	U4[i]	U5[i]	U6[i]	U7[i]	U[i]
1	2.7237	1.8574	2.7192	1.0685	2.5193	1.9217	1.8378	0.1374
2	1.7598	1.3374	1.3799	1.0729	1.6669	1.8514	1.7721	0.0708
3	2.1763	1.4644	2.3486	2.8179	2.7518	1.7316	1.5891	0.1946
4	2.8208	2.8552	1.1008	1.7019	1.3021	1.3523	1.7593	0.1517
5	1.7172	1.8404	2.1283	1.2743	1.1548	2.7849	2.9696	0.1406

6	1.6159	1.9890	2.8781	2.1846	1.3842	1.7660	2.3508	0.0853
7	2.9474	1.9001	2.5667	2.5755	2.4024	2.1875	2.3232	0.1506
8	2.8049	1.0100	1.4365	1.4722	1.8954	2.4615	1.6857	0.0655
9	1.2158	1.9758	1.3310	1.2437	1.7088	1.7046	1.7789	0.0264
10	2.9262	2.0368	2.2656	2.6516	1.8201	2.3484	1.4176	0.2500
...	...	...	...	...	...	...	...	...
491	2.1422	1.6907	2.8602	1.8178	2.3327	1.0667	2.1648	0.0498
492	2.3478	1.0168	1.9145	1.7788	1.5708	1.5813	2.2220	0.1438
493	2.8513	2.1335	2.6995	2.0881	2.0137	2.0042	1.9966	0.0200
494	1.0628	2.1577	1.6443	2.8320	2.0634	1.7277	2.7472	0.1831
495	1.2837	1.3890	2.5250	1.1436	2.7033	2.8526	2.3437	0.1528
496	1.5441	1.4641	2.9662	2.3198	1.0590	1.9024	2.2097	0.0377
497	1.0063	2.5562	2.4339	2.6643	2.0495	2.1898	2.0301	0.1152
498	2.6819	2.1866	1.6746	1.8651	2.1720	2.0051	2.1780	0.1565
499	2.8186	1.6034	1.5692	1.6257	2.1665	2.4986	2.0772	0.0705
500	2.0447	1.1046	1.4647	1.5830	1.7081	1.3174	2.0703	0.1634

Part B of program number 5.2 uses the random numbers of Table 5.4 according to acceptance-rejection sampling method and produces 7 sets of $n_1 = n_2 = n_3 = n_4 = n_5 = n_6 = n_7 = 6$ accepted random variates UA1, UA2, UA3, UA4, UA5, UA6 and UA7 for x_1, x_2, x_3, x_4, x_5, x_6, x_7 respectively shown in Table 5.5 for $N1 = 500$. Acceptance-rejection sampling technique uses the condition $U < norc\ (U1)(U2)(U3)(1/U4)(1/U5)(1/U6)(1/U7)$. If the condition is satisfied, the $U1$, $U2$, $U3$, $U4$, $U5$, $U6$ and $U7$ are accepted as random variates which we call $UA1[[i]]$, $UA2[[i]]$, $UA3[[i]]$, $UA4[[i]]$, $UA5[[i]]$, $UA6[[i]]$ and $UA7[[i]]$ respectively for further calculations.

Table 5.5 Showing accepted values of random variates obtained using part B of program number 5.2. We have 6 accepted values out of 500 for $U1$, $U2$, $U3$, $U4$, $U5$, $U6$ and $U7$.

i	UA1[[i]]	UA2[[i]]	UA3[[i]]	UA4[[i]]	UA5[[i]]	UA6[[i]]	UA7[[i]]
1	2.1373	2.8294	2.1605	2.7108	1.5260	2.5290	1.1250
2	2.1961	1.5937	1.8595	1.3074	2.0610	1.3927	2.3411
3	2.9062	2.4379	2.3912	1.9878	1.0104	1.9233	1.6296
4	2.6810	1.8672	2.4557	2.9314	1.4712	1.0822	1.2824
5	2.4114	2.7579	1.4093	2.0324	1.8773	1.4323	1.4719
6	2.9098	1.1156	1.7997	2.3364	1.8108	1.7853	1.5376

Using part C of program number 5.2, we evaluate the integral I_{14} using the following equation

$$I_{14} = \frac{1}{n_1\ n_2\ n_3\ n_4\ n_5\ n_6\ n_7} \sum_{i=1}^{n_1} \sum_{j=1}^{n_2} \sum_{k=1}^{n_3} \sum_{l=1}^{n_4} \sum_{m=1}^{n_5} \sum_{n=1}^{n_6} \sum_{o=1}^{n_7}$$

$$\frac{F(x_{1i}, x_{2j}, x_{3k}, x_{4l}, x_{5m}, x_{6n}, x_{7o})}{p(x_{1i}, x_{2j}, x_{3k}, x_{4l}, x_{5m}, x_{6n}, x_{7o})}$$

$$= \frac{1}{n_1\ n_2\ n_3\ n_4\ n_5\ n_6\ n_7} \sum_{i=1}^{n_1} \sum_{j=1}^{n_2} \sum_{k=1}^{n_3} \sum_{l=1}^{n_4} \sum_{m=1}^{n_5} \sum_{n=1}^{n_6} \sum_{o=1}^{n_7}$$

$$\frac{((x_{1i} + x_{2j} + x_{3k})/(x_{4l} + x_{5m} + x_{6n} + x_{7o}))^y}{norc\ (x_{1i})(x_{2j})(x_{3k})(1/x_{4l})(1/x_{5m})(1/x_{6n})(1/x_{7o})}$$

as

$$I_{14} = \frac{1}{n_1\ n_2\ n_3\ n_4\ n_5\ n_6\ n_7} \sum_{i=1}^{n_1} \sum_{j=1}^{n_2} \sum_{k=1}^{n_3} \sum_{l=1}^{n_4} \sum_{m=1}^{n_5} \sum_{n=1}^{n_6} \sum_{o=1}^{n_7}$$

$$\frac{(\ (UA1[[i]]+UA2[[j]]+UA3[[k]])/(UA4[[l]]+UA5[[m]]+UA6[[n]]+UA7[[o]])\ )^y}{norc\ \ (UA1[[i]])(UA2[[j]])(UA3[[k]])\ \ (1/UA4[[l]])(1/UA5[[m]])(1/UA6[[n]])(1/UA7[[o]])}$$

$$--------(5.6)$$

Results are in Table 5.6 as Monte Carlo result I_{mc}.

Using part C of program number 5.2, we have also calculated the integral as I_{ni} using NIntegrate command. Results are in Table 5.6 as I_{ni}. We have compared I_{ni} with I_{mc} for various values of parameters and of $N1$. The difference is denoted as error1 which is % error shown in Table 5.6.

Table 5.6 Results of part C of program number 5.2 for various values of parameters. We are dealing with

$$I_{14} = \int_a^b \int_a^b \int_a^b \int_a^b \int_a^b \int_a^b \int_a^b \left(\frac{x_1 + x_2 + x_3}{x_4 + x_5 + x_6 + x_7} \right)^y dx_1\ dx_2\ dx_3\ dx_4\ dx_5\ dx_6\ dx_7$$

i	y	a	b	$N1$	$n1$	$n1^7$	Monte Carlo result I_{mc}	result obtained using NIntegrate command I_{ni}	% error
1	1	1	3	500	6	279936	82.96	98.12	-15.46
2	1	1	3	650	8	2097152	86.57	98.12	-11.77
3	1	1	3	800	11	19487171	94.42	98.12	-3.77
4	2	1	3	600	8	2097152	75.08	79.12	-5.11
5	3	1	5	1500	8	2097152	9598.62	10319.70	-6.99
6	3	1	7	2000	5	78125	180883.00	199517.00	-9.34
7	3	1	7	2500	6	279936	197386.00	199517.00	-1.07

Table 5.6 displays a survey of Monte Carlo evaluation of the 7D definite integration I_{14} for different values of various parameters. We note that, while using Monte Carlo method, % error is generally below 10%. If the error is more than 10%, the error can usually be lowered by raising number of terms $n1^7$ in the summation in equation (5.6) by raising value of $N1$.

Chapter VI

Evaluation of Ten-dimensional Definite Integrals

This chapter deals with Monte Carlo evaluation of 2 ten-dimensional definite integrals. In each case, a suitable multi-variable probability density function is chosen and acceptance-rejection sampling is used to obtain values of 10 sets of random variates corresponding to the 10 random variables in the probability density function. The integrals are evaluated using the 10 sets of random variates. Programs written in Mathematica have been used in the sampling as well as in evaluating the integrals. Uses of different parts of the programs have been narrated.

6.1 The 10D definite integral dealt with in this chapter
This chapter deals with the following 2 definite integrals:

$$I_{15} = \int_a^b \int_a^b \int_a^b \int_a^b \int_a^b \int_a^b \int_a^b \int_a^b \int_a^b \int_a^b (x_1 + x_2 + x_3 + x_4 + x_5 + x_6 + x_7 + x_8 + x_9 + x_{10})^y$$

----------(6.1)

$$dx_1\ dx_2\ dx_3\ dx_4\ dx_5\ dx_6\ dx_7\ dx_8\ dx_9\ dx_{10}$$

$$I_{16} = \int_a^b \int_a^b \int_a^b \int_a^b \int_a^b \int_g^h \int_g^h \int_g^h \int_g^h \int_g^h (x_1 + x_2 + x_3 + x_4 + x_5 + x_6 + x_7 + x_8 + x_9 + x_{10})^y$$

----------(6.2)

$$dx_1\ dx_2\ dx_3\ dx_4\ dx_5\ dx_6\ dx_7\ dx_8\ dx_9\ dx_{10}$$

where y is positive constant such as 0.5, 1.5, 2, 2.5, 3 etc.; limits a and b are also constants such as 1, 5 etc.

6.2 Evaluation of the integral I_{15}
We have evaluated the multi-dimensional definite integral

$$I_{15} = \int_a^b \int_a^b \int_a^b \int_a^b \int_a^b \int_a^b \int_a^b \int_a^b \int_a^b \int_a^b (x_1 + x_2 + x_3 + x_4 + x_5 + x_6 + x_7 + x_8 + x_9 + x_{10})^y$$

$$dx_1\ dx_2\ dx_3\ dx_4\ dx_5\ dx_6\ dx_7\ dx_8\ dx_9\ dx_{10}$$

-------(6.3)

using acceptance-rejection sampling. Here $a = 1$, $b = 5$, $y = 0.5$, 1.5, 2, 2.5 and 3. We have written program number 6.1 in Mathematica using symbolic computation.

Program number 6.1
part A

```
y=2

N1=1000;

a=1;

b=5;

norc=1/(NIntegrate[x1*x2*x3*x4*x5*x6*x7*x8*x9*x10,

{x1,a,b},{x2,a,b},{x3,a,b},{x4,a,b},{x5,a,b},

{x6,a,b},{x7,a,b},{x8,a,b},{x9,a,b},{x10,a,b}])

c=norc*b*b*b*b*b*b*b*b*b*b

n=654321;

SeedRandom[n];


Table[{i=i+1,

U1[i]=a+(b-a)*RandomReal[],

U2[i]=a+(b-a)*RandomReal[],

U3[i]=a+(b-a)*RandomReal[],

U4[i]=a+(b-a)*RandomReal[],
```

```
U5[i]=a+(b-a)*RandomReal[],
U6[i]=a+(b-a)*RandomReal[],
U7[i]=a+(b-a)*RandomReal[],
U8[i]=a+(b-a)*RandomReal[],
U9[i]=a+(b-a)*RandomReal[],
U10[i]=a+(b-a)*RandomReal[],
u[i]=RandomReal[];U[i]=0+(c-0)*u[i]},{i,0,N1-1,1}];
TableForm[%,TableSpacing->{2,2},
TableHeadings->{None,{"i","U1[i]","U2[i]","U3[i]","U4[i]","U5[i]",
"U6[i]","U7[i]","U8[i]","U9[i]","U10[i]","U[i]"}}]
```

part B
```
UA1={};
i=0;
While[i<=N1,
If[U[i]<norc*U1[i]*U2[i]*U3[i]*U4[i]*U5[i]*U6[i]*U7[i]*U8[i]*U9[i]*U10[i],
AppendTo[UA1,U1[i]]];i=i+1];
n1=Length[UA1]


UA2={};
i=0;
While[i<=N1,
If[U[i]<norc*U1[i]*U2[i]*U3[i]*U4[i]*U5[i]*U6[i]*U7[i]*U8[i]*U9[i]*U10[i],
AppendTo[UA2,U2[i]]];i=i+1];
n2=Length[UA2]


UA3={};
i=0;
While[i<=N1,
If[U[i]<norc*U1[i]*U2[i]*U3[i]*U4[i]*U5[i]*U6[i]*U7[i]*U8[i]*U9[i]*U10[i],
AppendTo[UA3,U3[i]]];i=i+1];
n3=Length[UA3]


UA4={};
i=0;
While[i<=N1,
If[U[i]<norc*U1[i]*U2[i]*U3[i]*U4[i]*U5[i]*U6[i]*U7[i]*U8[i]*U9[i]*U10[i],
AppendTo[UA4,U4[i]]];i=i+1];
```

```
n4=Length[UA4]

UA5={};
i=0;
While[i<=N1,
If[U[i]<norc*U1[i]*U2[i]*U3[i]*U4[i]*U5[i]*U6[i]*U7[i]*U8[i]*U9[i]*U10[i],
AppendTo[UA5,U5[i]]];i=i+1];
n5=Length[UA5]

UA6={};
i=0;
While[i<=N1,
If[U[i]<norc*U1[i]*U2[i]*U3[i]*U4[i]*U5[i]*U6[i]*U7[i]*U8[i]*U9[i]*U10[i],
AppendTo[UA6,U6[i]]];i=i+1];
n6=Length[UA6]

UA7={};
i=0;
While[i<=N1,
If[U[i]<norc*U1[i]*U2[i]*U3[i]*U4[i]*U5[i]*U6[i]*U7[i]*U8[i]*U9[i]*U10[i],
AppendTo[UA7,U7[i]]];i=i+1];
n7=Length[UA7]

UA8={};
i=0;
While[i<=N1,
If[U[i]<norc*U1[i]*U2[i]*U3[i]*U4[i]*U5[i]*U6[i]*U7[i]*U8[i]*U9[i]*U10[i],
AppendTo[UA8,U8[i]]];i=i+1];
n8=Length[UA8]

UA9={};
i=0;
While[i<=N1,
If[U[i]<norc*U1[i]*U2[i]*U3[i]*U4[i]*U5[i]*U6[i]*U7[i]*U8[i]*U9[i]*U10[i],
AppendTo[UA9,U9[i]]];i=i+1];
n9=Length[UA9]

UA10={};
```

```
i=0;
While[i<=N1,
If[U[i]<norc*U1[i]*U2[i]*U3[i]*U4[i]*U5[i]*U6[i]*U7[i]*U8[i]*U9[i]*U10[i],
AppendTo[UA10,U10[i]]];i=i+1];
n10=Length[UA10]

Table[{i=i+1,
UA1[[i]],UA2[[i]],UA3[[i]],UA4[[i]],UA5[[i]],
UA6[[i]],UA7[[i]],UA8[[i]],UA9[[i]],UA10[[i]]},{i,0,n1-1,1}];
TableForm[%,TableSpacing->{2,2},
TableHeadings->{None,{"i","UA1[[i]]","UA2[[i]]","UA3[[i]]","UA4[[i]]",
"UA5[[i]]","UA6[[i]]","UA7[[i]]","UA8[[i]]","UA9[[i]]","UA10[[i]]"}}]
```

part C
```
Imc=(1/(n1*n2*n3*n4*n5*n6*n7*n8*n9*n10))*
Sum[((UA1[[i]]+UA2[[j]]+UA3[[k]]+UA4[[l]]+UA5[[m]]+
UA6[[n]]+UA7[[o]]+UA8[[p]]+UA9[[q]]+UA10[[r]])^y)/
(norc*UA1[[i]]*UA2[[j]]*UA3[[k]]*UA4[[l]]*UA5[[m]]*
UA6[[n]]*UA7[[o]]*UA8[[p]]*UA9[[q]]*UA10[[r]]),
{i,1,n1},{j,1,n2},{k,1,n3},{l,1,n4},{m,1,n5},
{n,1,n6},{o,1,n7},{p,1,n8},{q,1,n9},{r,1,n10}]

Ini=NIntegrate[(x1+x2+x3+x4+x5+x6+x7+x8+x9+x10)^y,
{x1,a,b},{x2,a,b},{x3,a,b},{x4,a,b},{x5,a,b},
{x6,a,b},{x7,a,b},{x8,a,b},{x9,a,b},{x10,a,b}]
error1=(Imc-Ini)*100/Ini
```

We have chosen a probability density function

$$p(x_1, x_2, x_3, x_4, x_5, x_6, x_7, x_8, x_9, x_{10}) = norc \ (x_1)(x_2)(x_3)(x_4)(x_5)(x_6)(x_7)(x_8)(x_9)(x_{10})$$

$$--------(6.4)$$

where *norc* is normalization constant. Both the integrand $(x_1 + x_2 + x_3 + x_4 + x_5 + x_6 + x_7 + x_8 + x_9 + x_{10})^y$ and $p(x_1, x_2, x_3, x_4, x_5, x_6, x_7, x_8, x_9, x_{10}) = norc \ (x_1)(x_2)(x_3)(x_4)(x_5)(x_6)(x_7)(x_8)(x_9)(x_{10})$ are increasing functions of all the 10 variables $x_1, x_2, x_3, \dots, x_{10}$. Hence maximum value of p is $c = norc \ (b)(b)(b)(b)(b)(b)(b)(b)(b)(b)$.

In part A of program number 6.1, $U1, U2, U3, \dots, U10$ are 10 sets of $N1 = N2 = N3 = \dots = N10$ uniform random numbers for $x_1, x_2, x_3, \dots, x_{10}$ respectively in the interval a to b. $U[i]$'s are $N1$ uniform random numbers in interval 0 to c.

Part B of the program uses the random numbers U_1, U_2, U_3, ..., U_{10} and U according to acceptance-rejection sampling method and produces 10 sets of n_1, n_2, n_3, ... , n_{10} accepted random variates $UA1$, $UA2$, $UA3$, ... , $UA10$ for x_1, x_2, x_3, ... , x_{10} shown in Table 6.1.

Acceptance-rejection sampling technique uses the condition $U < norc$ $(U1)(U2)(U3)(U4)(U5)(U6)(U7)(U8)(U9)(U10)$. If the condition is satisfied, the corresponding values of $U1$, $U2$, $U3$, ..., $U10$ are accepted as random variates which we call $UA1[[i]]$, $UA2[[i]]$, $UA3[[i]]$, ..., $UA10[[i]]$ respectively for further calculations.

Table 6.1 Showing accepted values of random variates $U1$, $U2$, $U3$, ... , $U10$ obtained using part B of program number 6.1. We have 5 accepted values out of 1000 values of $U1$, $U2$, $U3$, ... , $U10$.

i	UA1[[i]]	UA2[[i]]	UA3[[i]]	UA4[[i]]	UA5[[i]]	UA6[[i]]	UA7[[i]]	UA8[[i]]	UA9[[i]]	UA10[[i]]
1	4.8	4.4	2.7	3.3	4.7	3.3	4.4	2.1	4.1	1.3
2	5.0	3.5	3.6	1.5	3.7	2.2	4.7	1.9	4.2	4.4
3	3.8	4.2	2.3	2.0	2.1	4.5	3.1	4.7	1.5	3.8
4	3.8	2.1	3.0	4.5	2.9	4.0	2.5	2.5	4.5	4.7
5	3.9	2.5	4.1	4.9	3.5	3.2	2.5	4.2	1.3	4.8

Using part C of program number 6.1, we evaluate the integral I_{15} using the following equation

$$I_{15} = \frac{1}{n_1\,n_2\,n_3\,n_4\,n_5\,n_6\,n_7\,n_8\,n_9\,n_{10}} \sum_{i=1}^{n_1}\sum_{j=1}^{n_2}\sum_{k=1}^{n_3}\sum_{l=1}^{n_4}\sum_{m=1}^{n_5}\sum_{n=1}^{n_6}\sum_{o=1}^{n_7}\sum_{p=1}^{n_8}\sum_{q=1}^{n_9}\sum_{r=1}^{n_{10}}$$

$$\frac{F(x_{1i}, x_{2j}, x_{3k}, x_{4l}, x_{5m}, x_{6n}, x_{7o}, x_{8p}, x_{9q}, x_{10r})}{p(x_{1i}, x_{2j}, x_{3k}, x_{4l}, x_{5m}, x_{6n}, x_{7o}, x_{8p}, x_{9q}, x_{10r})}$$

$$= \frac{1}{n_1\,n_2\,n_3\,n_4\,n_5\,n_6\,n_7\,n_8\,n_9\,n_{10}} \sum_{i=1}^{n_1}\sum_{j=1}^{n_2}\sum_{k=1}^{n_3}\sum_{l=1}^{n_4}\sum_{m=1}^{n_5}\sum_{n=1}^{n_6}\sum_{o=1}^{n_7}\sum_{p=1}^{n_8}\sum_{q=1}^{n_9}\sum_{r=1}^{n_{10}}$$

$$\frac{(x_{1i} + x_{2j} + x_{3k} + x_{4l} + x_{5m} + x_{6n} + x_{7o} + x_{8p} + x_{9q} + x_{10r})^y}{norc\ (x_{1i})(x_{2j})(x_{3k})(x_{4l})(x_{5m})(x_{6n})(x_{7o})(x_{8p})(x_{9q})(x_{10r})}$$

as

$$I_{15} = \frac{1}{n_1\,n_2\,n_3\,n_4\,n_5\,n_6\,n_7\,n_8\,n_9\,n_{10}} \sum_{i=1}^{n_1}\sum_{j=1}^{n_2}\sum_{k=1}^{n_3}\sum_{l=1}^{n_4}\sum_{m=1}^{n_5}\sum_{n=1}^{n_6}\sum_{o=1}^{n_7}\sum_{p=1}^{n_8}\sum_{q=1}^{n_9}\sum_{r=1}^{n_{10}}$$

$$\frac{(\ UA1[[i]] + UA2[[j]] + UA3[[k]] + UA4[[l]] + UA5[[m]] + UA6[[n]] + UA7[[o]] + UA8[[p]] + UA9[[q]] + UA10[[r]])^y}{norc\ (UA1[[i]])\ (UA2[[j]])\ (UA3[[k]])\ (UA4[[l]])\ (UA5[[m]])}$$

$$(UA6[[n]])\ (UA7[[o]])(UA8[[p]])\ (UA9[[q]])(UA10[[r]])$$

--------(6.5)

Results are in Table 6.2 as Monte Carlo results I_{mc}.

Using part C of program number 6.1, we have also calculated the integral as I_{ni} using NIntegrate command. Results are in Table 6.2 as I_{ni}. We have compared I_{ni} with I_{mc} for various values of parameters. The difference has been shown as % error in Table 6.2.

Table 6.2 Results obtained using program number 6.1. Using $a = 1$, $b = 5$ for various values of parameters. We are dealing with

$$I_{15} = \int_a^b \int_a^b \int_a^b \int_a^b \int_a^b \int_a^b \int_a^b \int_a^b \int_a^b \int_a^b (x_1 + x_2 + x_3 + x_4 + x_5 + x_6 + x_7 + x_8 + x_9 + x_{10})^y$$

$$dx_1\ dx_2\ dx_3\ dx_4\ dx_5\ dx_6\ dx_7\ dx_8\ dx_9\ dx_{10}$$

y	$N1$	$n1$	$(n1)^{\wedge}10$	Monte Carlo result I_{mc}	result obtained using NIntegrate command I_{ni}	% error
0.5	1000	5	9765625	5.72155*10^6	5.7325*10^6	-0.2
1.5	1000	5	9765625	1.74861*10^8	1.73258*10^8	0.92
2	1000	5	9765625	9.70842*10^8	9.57699*10^8	1.37
2.5	1000	5	9765625	5.40546*10^9	5.31241*10^9	1.75
3	1000	5	9765625	3.0181*10^10	2.95698*10^10	2.1

6.3 Evaluation of the integral I_{16}

We have evaluated the multi-dimensional definite integral

$$I_{16} = \int_a^b \int_a^b \int_a^b \int_a^b \int_a^b \int_g^h \int_g^h \int_g^h \int_g^h \int_g^h (x_1 + x_2 + x_3 + x_4 + x_5 + x_6 + x_7 + x_8 + x_9 + x_{10})^y$$

$$dx_1\ dx_2\ dx_3\ dx_4\ dx_5\ dx_6\ dx_7\ dx_8\ dx_9\ dx_{10}$$

-------(6.6)

using acceptance-rejection sampling. Here $a = 1$, $b = 3$, $g = 1$, $h = 5$, $y = 0.25, 0.50, 0.75, 1.25, 1.50, 1.75, 2, 2.25, 2.50, 2.75$ and 3. We have written program number 6.2 in Mathematica using symbolic computation.

Program number 6.2
part A

```
y=3

N1=N2=N3=N4=N5=N6=N7=N8=N9=N10=600;

a=1;

b=3;

g=1;

h=5;

norc=1/(NIntegrate[x1*x2*x3*x4*x5*x6*x7*x8*x9*x10,

{x1,a,b},{x2,a,b},{x3,a,b},{x4,a,b},{x5,a,b},

{x6,g,h},{x7,g,h},{x8,g,h},{x9,g,h},{x10,g,h}])

c=norc*b*b*b*b*b*h*h*h*h*h

n=654321;

SeedRandom[n];


Table[{i=i+1,

U1[i]=a+(b-a)*RandomReal[],

U2[i]=a+(b-a)*RandomReal[],

U3[i]=a+(b-a)*RandomReal[],
```

85

```
U4[i]=a+(b-a)*RandomReal[],

U5[i]=a+(b-a)*RandomReal[],

U6[i]=g+(h-g)*RandomReal[],

U7[i]=g+(h-g)*RandomReal[],

U8[i]=g+(h-g)*RandomReal[],

U9[i]=g+(h-g)*RandomReal[],

U10[i]=g+(h-g)*RandomReal[],

U[i]=0+(c-0)*RandomReal[]},{i,0,N1-1,1}];

TableForm[%,TableSpacing->{2,2},

TableHeadings->{None,{"i","U1[i]","U2[i]","U3[i]","U4[i]","U5[i]",

"U6[i]","U7[i]","U8[i]","U9[i]","U10[i]","U[i]"}}]
```

part B
```
UA1={};

i=0;

While[i<=N1,

If[U[i]<norc*U1[i]*U2[i]*U3[i]*U4[i]*U5[i]*U6[i]*U7[i]*U8[i]*U9[i]*U10[i],

AppendTo[UA1,U1[i]]];i=i+1];

n1=Length[UA1]


UA2={};

i=0;

While[i<=N2,

If[U[i]<norc*U1[i]*U2[i]*U3[i]*U4[i]*U5[i]*U6[i]*U7[i]*U8[i]*U9[i]*U10[i],

AppendTo[UA2,U2[i]]];i=i+1];

n2=Length[UA2]


UA3={};

i=0;

While[i<=N3,

If[U[i]<norc*U1[i]*U2[i]*U3[i]*U4[i]*U5[i]*U6[i]*U7[i]*U8[i]*U9[i]*U10[i],

AppendTo[UA3,U3[i]]];i=i+1];

n3=Length[UA3]


UA4={};

i=0;

While[i<=N4,

If[U[i]<norc*U1[i]*U2[i]*U3[i]*U4[i]*U5[i]*U6[i]*U7[i]*U8[i]*U9[i]*U10[i],
```

```
AppendTo[UA4,U4[i]]];i=i+1];
n4=Length[UA4]


UA5={};
i=0;
While[i<=N5,
If[U[i]<norc*U1[i]*U2[i]*U3[i]*U4[i]*U5[i]*U6[i]*U7[i]*U8[i]*U9[i]*U10[i],
AppendTo[UA5,U5[i]]];i=i+1];
n5=Length[UA5]


UA6={};
i=0;
While[i<=N6,
If[U[i]<norc*U1[i]*U2[i]*U3[i]*U4[i]*U5[i]*U6[i]*U7[i]*U8[i]*U9[i]*U10[i],
AppendTo[UA6,U6[i]]];i=i+1];
n6=Length[UA6]


UA7={};
i=0;
While[i<=N7,
If[U[i]<norc*U1[i]*U2[i]*U3[i]*U4[i]*U5[i]*U6[i]*U7[i]*U8[i]*U9[i]*U10[i],
AppendTo[UA7,U7[i]]];i=i+1];
n7=Length[UA7]


UA8={};
i=0;
While[i<=N8,
If[U[i]<norc*U1[i]*U2[i]*U3[i]*U4[i]*U5[i]*U6[i]*U7[i]*U8[i]*U9[i]*U10[i],
AppendTo[UA8,U8[i]]];i=i+1];
n8=Length[UA8]


UA9={};
i=0;
While[i<=N9,
If[U[i]<norc*U1[i]*U2[i]*U3[i]*U4[i]*U5[i]*U6[i]*U7[i]*U8[i]*U9[i]*U10[i],
AppendTo[UA9,U9[i]]];i=i+1];
n9=Length[UA9]
```

```
UA10={ };

i=0;

While[i<=N10,

If[U[i]<norc*U1[i]*U2[i]*U3[i]*U4[i]*U5[i]*U6[i]*U7[i]*U8[i]*U9[i]*U10[i],

AppendTo[UA10,U10[i]]];i=i+1];

n10=Length[UA10]


Table[{i=i+1,UA1[[i]],UA2[[i]],UA3[[i]],UA4[[i]],UA5[[i]],

UA6[[i]],UA7[[i]],UA8[[i]],UA9[[i]],UA10[[i]]},{i,0,n1-1,1}];

TableForm[%,TableSpacing->{2,2},

TableHeadings->{None,{"i","UA1[[i]]","UA2[[i]]","UA3[[i]]","UA4[[i]]","UA5[[i]]",

"UA6[[i]]","UA7[[i]]","UA8[[i]]","UA9[[i]]","UA10[[i]]"}}]
```

part C

```
Imc=(1/(n1*n2*n3*n4*n5*n6*n7*n8*n9*n10))*

Sum[(((UA1[[i]]+UA2[[j]]+UA3[[k]]+UA4[[l]]+UA5[[m]]+

UA6[[n]]+UA7[[o]]+UA8[[p]]+UA9[[q]]+UA10[[r]])^y)/

(norc*UA1[[i]]*UA2[[j]]*UA3[[k]]*UA4[[l]]*UA5[[m]]*

UA6[[n]]*UA7[[o]]*UA8[[p]]*UA9[[q]]*UA10[[r]]),

{i,1,n1},{j,1,n2},{k,1,n3},{l,1,n4},{m,1,n5},

{n,1,n6},{o,1,n7},{p,1,n8},{q,1,n9},{r,1,n10}]


Ini=NIntegrate[(x1+x2+x3+x4+x5+x6+x7+x8+x9+x10)^y,

{x1,a,b},{x2,a,b},{x3,a,b},{x4,a,b},{x5,a,b},

{x6,g,h},{x7,g,h},{x8,g,h},{x9,g,h},{x10,g,h}]

error1=(Imc-Ini)*100/Ini
```

We have chosen a probability density function

$$p(x_1,x_2,x_3,x_4,x_5,x_6,x_7,x_8,x_9,x_{10}) = norc \ (x_1)(x_2)(x_3)(x_4)(x_5)(x_6)(x_7)(x_8)(x_9)(x_{10})$$

----(6.7)

where *norc* is normalization constant. Both the integrand $(x_1 + x_2 + x_3 + x_4 + x_5 + x_6 + x_7 + x_8 + x_9 + x_{10})^y$ and

$p(x_1,x_2,x_3,x_4,x_5,x_6,x_7,x_8,x_9,x_{10}) = norc \ (x_1)(x_2)(x_3)(x_4)(x_5)(x_6)(x_7)(x_8)(x_9)(x_{10})$ are increasing

functions of all the 10 variables $x_1, x_2, x_3, \ldots, x_{10}$. Hence maximum value of p is $c = norc \ (b)(b)(b)(b)(b)(h)(h)(h)(h)(h)$.

In part A of program number 6.2, $U1, U2, U3, \ldots, U10$ are 10 sets of $N1 = N2 = N3 = \ldots = N10$ uniform random numbers for $x_1, x_2, x_3, \ldots, x_{10}$ respectively in the interval a to b for x_1, x_2, x_3, x_4, x_5 and in the interval g to h for $x_6, x_7, x_8, x_9, x_{10}$. $U[i]$'s are $N1$ uniform random numbers in interval 0 to c.

Part B of program number 6.2 uses the random numbers $U_1, U_2, U_3, \ldots, U_{10}$ and U according to acceptance-rejection sampling method and produces 10 sets of $n_1, n_2, n_3, \ldots, n_{10}$ accepted random variates $UA1, UA2, UA3, \ldots, UA10$ for $x_1, x_2, x_3, \ldots, x_{10}$ shown in Table 6.3.

Acceptance-rejection sampling technique uses the condition $U < norc$ $(U1)(U2)(U3)(U4)(U5)(U6)(U7)(U8)(U9)(U10)$. If the condition is satisfied, the corresponding values of $U1$, $U2$, $U3$, ..., $U10$ are accepted as random variates which we call $UA1[[i]]$, $UA2[[i]]$, $UA3[[i]]$, ..., $UA10[[i]]$ respectively for further calculations.

Table 6.3 Showing accepted values of random variates $U1$, $U2$, $U3$, ... , $U10$ obtained using part B of program number 6.2. We have 4 accepted values out of 600 values of $U1$, $U2$, $U3$, ... , $U10$.

i	UA1[[i]]	UA2[[i]]	UA3[[i]]	UA4[[i]]	UA5[[i]]	UA6[[i]]	UA7[[i]]	UA8[[i]]	UA9[[i]]	UA10[[i]]
1	2.9	2.7	1.8	2.1	2.8	3.3	4.4	2.1	4.1	1.3
2	3.0	2.3	2.3	1.2	2.3	2.2	4.7	1.9	4.2	4.4
3	2.4	2.6	1.6	1.5	1.5	4.5	3.1	4.7	1.5	3.8
4	2.4	1.5	2.0	2.8	2.0	4.0	2.5	2.5	4.5	4.7

Using part C of program number 6.2, we evaluate the integral I_{16} using the following equation

$$I_{16} = \frac{1}{n_1\, n_2\, n_3\, n_4\, n_5\, n_6\, n_7\, n_8\, n_9\, n_{10}} \sum_{i=1}^{n_1} \sum_{j=1}^{n_2} \sum_{k=1}^{n_3} \sum_{l=1}^{n_4} \sum_{m=1}^{n_5} \sum_{n=1}^{n_6} \sum_{o=1}^{n_7} \sum_{p=1}^{n_8} \sum_{q=1}^{n_9} \sum_{r=1}^{n_{10}}$$

$$\frac{F(x_{1i},x_{2j},x_{3k},x_{4l},x_{5m},x_{6n},x_{7o},x_{8p},x_{9q},x_{10r})}{p(x_{1i},x_{2j},x_{3k},x_{4l},x_{5m},x_{6n},x_{7o},x_{8p},x_{9q},x_{10r})}$$

$$= \frac{1}{n_1\, n_2\, n_3\, n_4\, n_5\, n_6\, n_7\, n_8\, n_9\, n_{10}} \sum_{i=1}^{n_1} \sum_{j=1}^{n_2} \sum_{k=1}^{n_3} \sum_{l=1}^{n_4} \sum_{m=1}^{n_5} \sum_{n=1}^{n_6} \sum_{o=1}^{n_7} \sum_{p=1}^{n_8} \sum_{q=1}^{n_9} \sum_{r=1}^{n_{10}}$$

$$\frac{(x_{1i}+x_{2j}+x_{3k}+x_{4l}+x_{5m}+x_{6n}+x_{7o}+x_{8p}+x_{9q}+x_{10r})^y}{norc\ (x_{1i})(x_{2j})(x_{3k})(x_{4l})(x_{5m})(x_{6n})(x_{7o})(x_{8p})(x_{9q})(x_{10r})}$$

as

$$I_{16} = \frac{1}{n_1\, n_2\, n_3\, n_4\, n_5\, n_6\, n_7\, n_8\, n_9\, n_{10}} \sum_{i=1}^{n_1} \sum_{j=1}^{n_2} \sum_{k=1}^{n_3} \sum_{l=1}^{n_4} \sum_{m=1}^{n_5} \sum_{n=1}^{n_6} \sum_{o=1}^{n_7} \sum_{p=1}^{n_8} \sum_{q=1}^{n_9} \sum_{r=1}^{n_{10}}$$

$$\frac{(\ UA1[[i]]+UA2[[j]]+UA3[[k]]+UA4[[l]]+UA5[[m]]+ UA6[[n]]+UA7[[o]]+UA8[[p]]+UA9[[q]]+UA10[[r]])^y}{norc\ (UA1[[i]])\ (UA2[[j]])\ (UA3[[k]])\ (UA4[[l]])\ (UA5[[m]])}$$

$$(UA6[[n]])\ (UA7[[o]])(UA8[[p]])\ (UA9[[q]])(UA10[[r]])$$

--------(6.8)

Results are in Table 6.4 as Monte Carlo results I_{mc}.

In part C of the program, we have also calculated the integral as I_{ni} using NIntegrate command. Results are in Table 6.4 as I_{ni}. We have compared I_{ni} with I_{mc} for various values of parameters. The difference has been shown as % error in Table 6.4.

Table 6.4 Results obtained using program number 6.2. Using $a = 1, b = 3, g = 1, h = 5$ for various values of parameters. We are dealing with

$$I_{16} = \int_a^b \int_a^b \int_a^b \int_a^b \int_a^b \int_g^h \int_g^h \int_g^h \int_g^h \int_g^h (x_1 + x_2 + x_3 + x_4 + x_5 + x_6 + x_7 + x_8 + x_9 + x_{10})^y$$

$$dx_1\ dx_2\ dx_3\ dx_4\ dx_5\ dx_6\ dx_7\ dx_8\ dx_9\ dx_{10}$$

y	$N1$	$n1$	$(n1)^{\wedge}10$	Monte Carlo result I_{mc}	result obtained using NIntegrate command I_{ni}	% error
3	600	4	1048576	5.09693*10^8	5.3248*10^8	-4.279
2.75	600	4	1048576	2.25828*10^8	2.36315*10^8	-4.437
2.50	600	4	1048576	1.00128*10^8	1.04958*10^8	-4.609
2.25	600	4	1048576	4.44265*10^7	4.66528*10^7	-4.772
2	600	4	1048576	1.97262*10^7	2.07531*10^7	-4.948
1.75	600	4	1048576	8.76512*10^6	9.23916*10^6	-5.130
1.50	600	4	1048576	3.89753*10^6	4.11653*10^6	-5.319
1.25	600	4	1048576	1.73436*10^6	1.83562*10^6	-5.516
0.75	600	4	1048576	344193.	365895.	-5.931
0.50	600	4	1048576	153504.	163564.	-6.150
0.25	600	4	1048576	68510.9	73178.4	-6.378

Concluding remarks

1) We have evaluated a number of multi-dimensional definite integrals using Monte Carlo method.

2) These are 2, 3, 5, 7 and 10 dimensional definite integrals.

3) We have chosen suitable *multi-variable* probability density functions and obtained corresponding random variates using the random sampling method called *acceptance-rejection* method.

4) We have used probability density functions that go like the integrands; i.e. if the integrand F is an increasing function of a variable, the probability density function p is chosen to be an increasing function of the variable. If the integrand is a decreasing function of a variable, the probability density function is chosen to be a decreasing function of the variable.

5) We have shown that F and p need not be strictly proportional to each other. Satisfactory results can be achieved if p follows F i.e. if both are increasing function of a variable or if both are decreasing function of a variable.

6) We have performed symbolic computations using programs written in Mathematica®. Hence the programs are evident to course conductors if not to students.

7) We have performed sampling of random variates as well as Monte Carlo integrations using the programs.

8) We have also evaluated the same definite integrals using *NIntegrate* command of Mathematica.

9) % error between results obtained using Monte Carlo method and those obtained using *NIntegrate* command of Mathematica are obtained as check of Monte Carlo results.

10) Necessary introduction is provided in chapter 1 so that the book is *self-contained*.

11) The book will contribute its mite in further evaluation of multi-dimensional Monte Carlo Integrations.

Reference

[1]   Istvan Manno; Introduction to the Monte-Carlo Method
       Akademiai Kiado, Budapest, Hungary (1999)
       page number 62

For further studies:
Books by the author (Sujaul Chowdhury *et al.*) to serve as course books for the course and lab. titled *Computational Physics* or *Computational Mathematics*

1) Computational Physics
   American Academic Press (2021)

2) Numerical Solutions of Initial Value Problems Using Mathematica
   IOP Concise Physics (2018)

3) Numerical Exploration of Fourier Transform and Fourier Series: The Power Spectrum of Driven Damped Oscillators
   Springer (2023)

4) Numerical Solutions of Boundary Value Problems with Finite Difference Method
   IOP Concise Physics (2018)

5) Numerical Solutions of Boundary Value Problems with Derivative Boundary Conditions
   American Academic Press (2019)

6) Numerical Solutions of Boundary Value Problems with So-called Shooting Method
   Nova Science Pub Inc (2021)

7) Numerical Solutions of Boundary Value Problems of Non-linear Differential Equations
   CRC Press (2021)

8) Numerical Solutions of Partial Differential Equations Using Finite Difference Method And Mathematica
   American Academic Press (2019)

9) Monte Carlo Methods: A Hands-On Computational Introduction Utilizing Excel
   Springer (2021)

10) Monte Carlo Methods Utilizing Mathematica®: Applications in Inverse Transform and Acceptance-Rejection Sampling
   Springer (2023)

11) Multi-dimensional Monte Carlo Integrations Utilizing Mathematica®
   American Academic Press (2025)